ONE UNION IN WOOD

JERRY LEMBCKE
and
WILLIAM M. TATTAM

Published simultaneously in Canada and the U.S. by
HARBOUR PUBLISHING CO. LTD.
and
INTERNATIONAL PUBLISHERS

ONE UNION IN WOOD

PUBLISHED BY

Harbour Publishing
Box 219 Madeira Park
British Columbia V0N 2H0 (Canada)

International Publishers
381 Park Avenue South
New York, NY 10016 (U.S.A.)

Cover design by Dianne Bersea
Typesetting and layout by Baseline Type & Graphics Cooperative,
  a worker owned and operated cooperative
Printed in Canada

Portions of this text have appeared in *Political Power and Social Theory, Labour/Le Travailleur,* and *Labour History.*

CANADIAN CATALOGUING IN PUBLICATION DATA

Lembcke, Jerry.
  One union in wood

Includes Index.
Bibliography: p.
ISBN 0-920080-43-X

1. International Woodworkers of America — History.
I. Tattam, William.   II. Title
HD6515.W6L45 1983   331.88'1349   C83-091343-2

LIBRARY OF CONGRESS CATALOGUING IN PUBLICATION DATA

Lembcke, Jerry, 1943-
  One union in wood

  Bibliography: p.
  Includes index.
    1. International Woodworkers of America — History.
  2. Trade-unions — Woodworkers — United States — History.
  3. Trade-unions — Woodworkers — Canada — History.
  I. Tattam, William, 1941-   II. Title.
  HD6515.W6L45 1984   331.88'194'0973   84-15808
  ISBN 0-7178-0619-7

Table of Contents

Foreword

One Union in Wood is the most accurate account yet of the effort to organize a democratic industrial union in the North American wood-products industry. The authors have refuted the conclusions of Vernon Jensen and other writers that class collaboration is the natural state of affairs in industry. In its place the authors have shown that the age old struggle between labour and capital continues and that the capital class knows no national or cultural borders in its pursuit of profit.

From the early formation of the IWA, the chosen and democratically elected leadership of this great union fully understood the forces determining their position in society. For them, the union struggle was only a part of a larger political struggle waged between labour and capital. The so-called White Bloc minority, on the other hand, could only embrace the struggle for wages, hours and working conditions which led them up a blind alley. The rank-and-file workers in the woods and mills were quite aware of the differences in leadership and supported those elected in the union's early years. The White Bloc was only able to come to power with the assistance of an anti-communist campaign waged by the governments of Canada and the United States and the employers on both sides of the border.

The history set out in these pages is a valuable contribution to the ongoing efforts to make the IWA a militant and democratic union although its general analysis and conclusions are applicable to the whole labour movement. The authors have not presumed to give all the answers but they have provided a new starting point for research and writing on the IWA.

Harold Pritchett
1st International President, IWA

Introduction

For over forty years, "One Union in Wood" has been the organizing slogan for workers in North America's forest products industry. This slogan encapsulates the attempts of loggers and lumberjacks, boommen and rafters, sawmill and plywood workers to build a democratic industrial union. In addition, "one union in wood" reflects the aspirations of woodworkers facing the powerful political and economic forces which frustrate and fragment their efforts to unionize.

Woodworkers have traditionally suffered the most dangerous and degrading working conditions found in North American industry; death and dismemberment routinely visited the woods; cold, damp, lice-infested bunk houses were "home" to loggers; holidays, health benefits and pensions were non-existent. Sixty-hour work weeks were common. And when woodworkers tried to organize, employers met them with blacklists and lock-outs.

In 1937 woodworkers rebelled against these conditions and severed their ties to the Sawmill and Timberworkers Union, an affiliate of the American Federation of Labor's United Brotherhood of Carpenters and Joiners. In its place they formed the International Woodworkers of America (IWA). It would become the largest affiliate of the Congress of Industrial Organizations (CIO) and the Canadian Congress of Labour (CCL) on the West Coast.

Given its importance and colorful IWW heritage, the IWA has been surprisingly under-researched.[1] The first and most widely known study was *Lumber and Labor* (1945) by Vernon Jensen. Irving Bernstein and Walter Galenson devoted brief sections of their books to the formation of the union. In addition, Paul Phillips, Irving Abella and Gad Horowitz completed short pieces on the IWA in Western Canada. Grant MacNeil and the IWA's International office have produced "official" versions of IWA history. More recently, Bert Cocheran and Harvey Levenstein have relied upon IWA history to make points crucial to their work on communism and trade unions, but historians have largely ignored the period from 1942 to 1959, and no one to date has attempted a *complete* history of the union.

Vernon Jensen's history of the IWA, published in 1945 and written while he was arbitrating a major dispute between the union and employers, has remained the definitive historical work on the union, and has, therefore, been an important influence on a generation of CIO scholarship. His work however, covered only the first four years of the union's history. While those four years may, in some sense, constitute the formative period, events more important to our present understanding occurred in the cold war years and well beyond. Moreover, his conclusion that native rank-and-file anti-communism crystalized in 1941 to oust, by democratic means, the union's Communist leadership does not hold up under reexamination. Jensen found that the "average timberworker" had "rejected Communism" and "changed the leadership of the IWA using democratic procedures." He called the episode "trade union democracy at work."[2] Jensen's case study of the IWA premised itself on the conservatism of rank-and-file workers and the neutrality of the CIO's National Office vis-a-vis internal union struggles.

One Union in Wood reconsiders the case of the IWA, arguing that the divisions which beset it were more than factional power struggles, and that the political equations decisive for the union's future were more complex than the simple rank-and-file anti-communism portrayed by Jensen. This book begins instead with the premise that the inherent contradiction in the exploitation of labor by capital is the mainspring of history. The first chapter details the early social history of the industry because it is in the industry's uneven historical development that we find the roots of the factions that vied for control of the IWA. Chapters two and three focus on the economic social conditions of the middle 1930s and the seminal events leading to the formation of the IWA: the 1935 lumber strike, the rise of anti-communism and local government repression as a reaction to industrial unionism, and the woodworkers' break with the American Federation of Labor (AFL). Chapters 4 through 6 show that the decisive interventions against the IWA's Communist leaders were made, not by the rank and file, but by the U.S. and Canadian governments, the national offices of the Congress of Industrial Organizations (CIO) and the Canadian Congress of Labour (CCL), and that the means used were by no measure democratic.

The most important of the state interventions were the deportation of the

International's first president in 1940, the denial to Communists of free border passage, and the passage of the Taft-Hartley Law's anti-communist provisions and their application to Canadian as well as American* IWA members. During the 1940s the leadership ranks and organizational structures of industrial unions like the IWA were breached by influences that diluted their democratic working-class character. The decisive actors in the IWA case were professional trade unionists like Adolph Germer, who was assigned by the CIO national office to take over the union's organizing program. Germer, through his experience as a factionalist in other CIO unions and in the Socialist Party, provided the organizational expertise and ideological sustenance to conservatives within the IWA, while acting as a conduit for the resources that flowed from the national office to fight the union's left wing. In the late 1940s the CIO combined forces with the Canadian Congress of Labour (CCL) and the Cooperative Commonwealth Federation (CCF) to force a secession of the left-led British Columbia district council from the IWA. In regard to the Canadian case, we argue, contra Irving Abella, that it was not rank-and-file volition and errors made by the B.C. Communists that undid the left, but rather that it was the resourceful combination of corporate and state power with social democratic trade unionists that held the key.[3]

The convergence of capitalist class interests with the interests of a trade union leadership faction whose class identity was ambiguous became the vehicle for capital's penetration of working class organizations. It is this intricate alignment of class forces at a time when the American-Canadian political economy was being restructured by post-World War II developments that explains the abrupt change of course taken by North American industrial unions in the late 1940s and 1950s. By 1958, business union leadership had consolidated its hold on the IWA, and the final chapter of this book portrays the loss of critical strikes in Newfoundland and Laurel, Mississippi as a legacy of this consolidation.

The theoretical implication of this reexamination is that the political make-up of union leadership at any historical moment is determined by the intersection of struggles internal to the working class with the political and economic struggles external to it. What appeared to be personal and ideological differences between union leaders and factions within the CIO-CCL unions were in reality struggles between strata of the working class whose historic roots were deeply embedded in the uneven development of North American capitalism. Implicit in this analysis is the recognition that to

---

* The authors are aware that the use of "America" to refer to the "United States" is controversial. We have, however, deferred to the editor's preference for the colloquial form.

understand the CIO-CCL period, additional research work must be done within a conceptual framework very different from the earlier histories. We have attempted to take a first step toward such a framework in *One Union in Wood*.

By emphasizing the political history of the IWA, other historical and sociological aspects of the union may have been slighted. A more extensive social history of the industry, for example, would add a great deal to present understanding. We know that immigrants brought rich cultural and political traditions to the woods, but we do not fully understand the link between their time and ours. The relationships between technological development, division of labor, and class struggle at the work place also await a full exploration, and a veritable "gold mine" awaits the researcher who approaches the wood products industry with this conceptual framework. While we regret that we have not been able to examine all these areas ourselves, we heartily encourage others to do so.

The authors alone are responsible for the theoretical and methodological choices and the interpretations made of their data. There are, however, many individuals and organizations who have supported our work and to whom we are grateful. Our greatest debt is to several people who helped the IWA through its first forty years and who made invaluable contributions to the research and writing of this book. They include (in alphabetical order) Ed Benedict, Bruce Bishop, Bill Botkin, Don Downing, Frank Fuller, William Harris, Ed Kenney, Harvey Ladd, Ralph Nelson, Harold Pritchett, Ron Roley, and Lyman Wax. Karly Larsen, Ralph Nelson and Harold Pritchett were particularly helpful and generous with their time and materials.

The authors have relied upon primary documents only recently made available to reinterpret earlier findings and explore wholly new historical areas. The Harold Pritchett papers at both the University of British Columbia and the University of Washington were especially useful to establish the presence of indigenous rank-and-file leftism and to complete the story of Pritchett's deportation case. The Adolph Germer Papers at the Wisconsin State Historical Society and the record of Adolph Germer's "trial," found in the IWA's International office, were valuable in documenting the disruptive role played by Germer. The "Minutes and Proceedings" of the Federation of Woodworkers and IWA proved indispensable. Collections of personal papers made available by Karly Larsen, Ralph Nelson, William Harris and Nina Hartung have also been very helpful. All the former IWA International Presidents who were alive during the period of research were interviewed, except for one who refused. Numerous other former officers, staff persons, members and political activists made invaluable contributions to the development of the book.

To those who are continuing to build the IWA, we are grateful for their assistance and encouragement. The research and education staff in the International office — Bob Baugh, Bettina Canon, Joyce Johnson, Roy Ockert and Denny Scott — took time out from their busy schedules to help locate materials and discuss ideas. The International officers during the time of writing — Keith Johnson, Bob Gerwig, Bud Rahberger and Fernie Viala — were patient and cooperative when it would have been easy for them to be otherwise. Many regional officers and staff persons, especially Clay Perry, John Hutter, and Charlie Campbell, helped develop regional material.

The section on the South benefited from the insights of Jim Youngdahl and the archival work of Bob Dinwiddie at the Southern Labor Archives at Georgia State University. In St. John's, Newfoundland, Donna Butt, Ric Boland, James Green, Q.C. and Professor Rolf Hattenhauer, Memorial University, as well as numerous lumber workers mentioned in the footnotes are due thanks for their cooperation. The Regional IWA office in Toronto, Canada, headed by Jean-Marie Bedard, and the collection of "IWA Newfoundland Papers" found there were extremely helpful. The Midwest section could not have been completed without the assistance of Debra Bernhardt, Irene Paull, Ernest Tomberg, and Oliver Rasmussen.

Many people, in different ways, have lent us personal encouragement. For Bill: My work began as a master's thesis in American History at the University of Oregon in 1968 where advising Professor Edwin Bingham offered support. As a doctoral candidate in sociology at the University in 1976 Jerry Lembcke conceptualized the project and kept it going. My father, Thomas M. Tattam, before his death, and my mother, Marjorie G. Tattam, both English immigrants, nurtured my contribution in innumerable ways. My wife Sara, enhanced our work considerably with her editing skills, and with our sons, Tom and Ian, tolerated my many absences with patience and understanding.

For Jerry: Debts are owed to friends and fellow workers in Eugene, Oregon, who helped me through my graduate program and from whom I learned a great deal. Members of the *Insurgent Sociologist* and Pacific Northwest Research Center collectives and friends in the Graduate Teaching Fellows Federation were especially supportive. Special thanks are due to Paul Fitzgerald, Marty Hart-Landsberg, Carolyn Howe, Bob Marotto and Al Szymanski and, of course, Bill Tattam for their encouragement and companionship over the years.

1
Uneven Development: The Structural Basis of Union Factionalism

Lumbering: An Early Migratory Industry

When North America's first European settlers began hacking their survival from the forests of the eastern seaboard, there appeared to be no end to the stands of giant timber. These stands provided the pit-sawn lumber from which the settlers built their wood-frame houses and the timber which made shipbuilding one of the continent's first manufacturing industries. The trees even provided the cargoes of "clapborde" which became one of the new land's first exports.

But as the trees along the coast were cut, the cutting teams had to move inland. By the early 1700s, it had become necessary for them to go at least twenty miles inland to find mast-sized timber, and then float the logs down the rivers; in this way, 120-foot-long pines made the trip down to the coast to become masts for the ships of the Royal Navy.[1] On the St. John and the Penobscot Rivers of New Brunswick, log rafting techniques were developed and perfected.

By the time of the Revolutionary War, the demand for ships' masts and hulls, casks and barrels had spread the industry throughout the North Atlantic colonies. At the same time the industry's reputation for frontier-type living conditions was also being established. Henry David Thoreau wrote that the hovels of New England woodsmen and those of their cattle were hardly distinguishable, except that the latter had no chimney. Working hours for these early woodworkers went from dawn to dark for wages of approximately $35 per month.[2] In the evenings, the men sat around their smoke-filled cabins in sub-zero temperatures.

There were no bunks and no chairs. The men ate standing, out of a common

1

kettle. For a bed, fir and hemlock boughs were sometimes strewn on the bare ground and on these was laid a twenty-foot wide spread stuffed six inches thick with cotton batting. When one of those things got wet, twenty men could hardly lift it. Such spreads were still being used on drive(s) as late as 1930. On top of the first one was spread a second, and about a dozen men crawled in between, lying spoon-fashion.....Old woodsmen aver that the worst thing about them was when some man would crepitate underneath them. And of course, the inevitable lice found them a favorite roosting place.[3]

The Midwest and South

The forests of the eastern seaboard were depleted early in the 1800s, but the construction of the Erie Canal in 1817 had opened the Michigan woods for logging. After the Civil War, the growth of the railroads and the demand for construction lumber in the prairie states spread the industry into Wisconsin and Minnesota. As a consequence, the entrepreneurs and workers, with their respective accumulations of capital, skills and knowledge, left the maritimes and followed the industry westward.[4] By 1880, large numbers of German, Scandinavian and Canadian immigrants had arrived in the Lake States, many of them single men carrying all their possessions in "turkeys" or "balloons"[5] which they carried on their backs. "Home" for them was the logging camp shanty, a one-room log structure with little or no ventilation, which quickly filled with smoke, steam and the odor "from drying unwashed socks, mittens, and clothes hung around and above the central stove."[6]

The logger's day began "any time after midnight." Winter, when the ground was frozen and the logs could be skidded out of the forest on the snow, was the prime time for logging in the Lake States.

> About eleven o'clock in the morning, the noon meal would be delivered to the woods on a sleigh....A crackling good fire would be burning when the hungry men gathered. Nevertheless, the beans would sometimes freeze to the tin plates in spite of the roaring fire. The slabs of pork and the biscuits, on the point of freezing before they were devoured, gave warmth and energy within. Afterward, it was on the job again until dusk. Then came the night meal back in camp.[7]

It was in the Lake States that Frederick Weyerhaeuser and his partner, F.C.A. Denkmann got their start. Investing in a small mill in Rock Island, Illinois, they purchased logs cut on the upriver tributaries of the Mississippi; soon they acquired a second mill and their own timber acreage. Their logging methods were the same as those which had been established earlier in the Northeast — cut and run. Thus, a robber baron in lumber to compliment those in steel and railroading was on his way.[8]

By the late 1800s, Frederick Weyerhaeuser was traveling as far north as the Chippewa River in northwestern Wisconsin in search of suitable trees for his Rock Island sawmill.[9] He merged his operation with several other companies and acquired more land. When the Wisconsin forests were cut over, he turned to Minnesota, formed more new companies and bought still more land.

But the Lake States' lumber supply was not inexhaustible. The timber barons had cut fifty-seven million acres of forest in forty years — the most complete and rapid cutting of so large a forest to that time. In their wake, they left "depleted timber resources, unused railroads, and stranded towns" and a resultant "economic depletion and impoverishment of a vast area and the veritable decay of many hard-working people."[10]

With the depletion of his Midwestern resources, Weyerhaeuser went west, where his company would become one of North America's largest multi-national corporations. On January 3, 1900, he and his associates purchased 900,000 acres of Washington State timber land at $6 per acre from railroad magnate James Hill.[11]

The South held no attraction for Weyerhaeuser: "the Negroes were too lazy," he is reported to have said. Other lumber capitalists did like the South, however.[12] The Southern Homestead Act of 1866 which had prevented the acquisition of large tracts of timber was repealed in 1876, opening the way for large-scale exploitation of timber reserves in Mississippi, Louisiana and East Texas. In the yellow pine country of the Gulf States, the cutting was as rapid as it had been in the Lake States, reaching twenty billion board feet in 1899.

Operators found a ready supply of cheap labor in the South. Black workers were commonly employed in the large operations while whites were more numerous in the small operations scattered throughout the countryside. The southern work force was not migratory; many of the workers shifted from lumber to agriculture, eking out an existence on small plots of land, allowing them to stay in one location despite mill closures or cyclical fluctuations in lumber production. In any case, mobility for these workers was difficult. Company stores, where the workers were often forced to shop, made extensive use of scrip, coupons and other cash substitutes; lacking cash, the workers became indebted to their employer for monthly loans or draws. Strict vagrancy laws further checked the movement of workers; arrested for vagrancy, a worker would be bailed out by an employer only to face a debt to be repaid through more labor.

By the 1920s, many areas of the South had been cut over but, unlike the lumber industry of the northeast and the Lake States, the southern lumber industry did not join the westward migration because the fast-growing yellow pine promised another generation of trees within a lifetime. In the meantime, the mill villages were closed, guaranteeing endemic poverty for the workers who remained, as most southern workers did, for the paternalism of class relations held the work force to the land. A half-century later, following the technological and political maturation of the industry in the Douglas fir region of the Pacific Northwest, the industry would return to the South to renew its exploitation of both forests and workers.

The Pacific Northwest and British Columbia

The earliest lumber mills in the Pacific Northwest were located along the lower Columbia River and along the lower reaches of the Willamette River near what is now Portland, Oregon. The earliest of these was probably established in 1827 by the Hudson's Bay Company about six miles up the Columbia River from Fort Vancouver.[13] Additional mills were established near Willamette Falls (now Oregon City) and on the Tualatin Plains by the late 1840s, as more and more settlers of "comfortable means" arrived and the demand for lumber for local construction increased.[14]

Small mills also sprang up along the Oregon coast to process the huge spruce and hemlock which could be floated downstream to them. The shipment of lumber and logs along the coast, especially to San Francisco, provided a comfortable profit for these mills until the late 1890s, but by that time the industry had become dominated by a few large firms on Puget Sound which were backed by capital from California and the East. While some of the mills on the Oregon Coast were able to continue to cut for local building construction, many succumbed after deep-water steamship transportation began to serve the larger mills on Puget Sound.

These big operations had been attracted by the abundant timber and deep harbors in the area around Seattle. The Pope and Talbot Company established one of the most successful operations, the Puget Mill Company, at Port Gamble, Washington, in 1852. Built as a replica of the New England town where the Pope and Talbot company headquarters was located, Port Gamble epitomized the company mill town.[15] In 1858 the company built a second mill, "the biggest in the West," and within twenty years nearly dominated the Pacific Northwest, although other companies had established large mills in the Grays Harbor (Aberdeen, Washington) area.[16]

The industry's later start in Washington ensured that its development would be qualitatively different from that in Oregon. In 1880, when Washington's lumber industry had only a fraction of the number of mills in Oregon, its mills had fifty-five percent more capital (see Table I). The arrival of Weyerhaeuser twenty years later would insure that the capitalization of the Washington mills would remain high for decades.

The wood products industry got off to a later and less promising start in British Columbia. A British monopoly established the Alberni Mill company in the 1860s, but although in time it employed 700 workers, U.S. trade barriers kept the firm's lumber out of San Francisco, leaving it with insufficient markets. It was forced to close before 1870.[17] Not until the railroads punched west through the rugged Canadian Rockies in the late 1880s and opened the interior of British Columbia did the industry boom. By 1899 B.C. lumber production had increased to 252,580 million board feet. Nevertheless, British Columbia's output was only one-third that of Oregon and one-sixth that of Washington. (Table II)

This uneven regional development was attended by two factors which shaped the political development of the industry's workers. First was the settlement pattern that brought conservative agrarian workers to the rich farmland of the lower Columbia River in the mid-nineteenth century, while in the last decades of the nineteenth century workers with industrial experience and socialist leanings were drawn to the big mills in Northern Washington and British Columbia. Second was the uneven development of

TABLE I.   **Establishments and Capital in Oregon and Washington Lumber Industry, 1880**

	Number of Establishments	Capital	Capital/Number of Establishments
Oregon	228	$1,577,875	6,920
Washington	37	$2,456,450	66,390

Source: Maxwell Reprint Company, American Industry and Manufacturers in the 19th Century, Vol. 8. Maxwell Reprint Company, Elmsford, New York.

TABLE II.   Lumber Production in Oregon,
                    Washington, and British Columbia,
                      1849-1899 (million board feet)

Year	Oregon	Washington	British Columbia
1849	16,853	4,080	— —
1859	41,169	77,125	1,750
1869	75,193	128,743	25,000
1879	177,171	160,176	50,000
1889	444,565	1,061,560	67,612
1899	734,181	1,428,205	252,580

Source: Thomas R. Cox, *Mills and Markets: A History of the Pacific Coast Lumber Industry to 1990* (Seattle: University of Washington Press, 1974), p. 301.

unionization. In the lower Columbia River area around Portland, craft unionism had gained a foothold in the 1880s, while from the outset workers farther north, especially in British Columbia, formed more advanced industrial unions.

Social Cleavages in the Workforce

The immigrants' contribution to the political development of North American unionism has long been in dispute. Some historians have argued that the church-dominated peasant backgrounds of the immigrants, coupled with their drive to "make it" and return to the "old country" with their savings, made them a force for conservatism in the North American labor movement. Others have noted that when the immigrants with prior European trade union and socialist political experience confronted the harsh conditions here, they were likely to become a radical element in the working class.[18]

The role of immigrants in the woods of the Pacific Northwest and British Columbia is no less complex. Their influence varied depending upon the nature of the experience they brought with them and the time of their arrival. Oregon had the earliest immigration, but both Washington and British Columbia soon surpassed it in numbers. Oregon and Washington reached peak percentages of immigration in the last decade of the nineteenth century, while British Columbia continued to climb to an all-time high in 1920 when fifty-three percent of the provincial population was foreign-born (Table III). In 1910 foreign-born workers made up nearly fifty percent of the laborers in Pacific Northwest mills but their numbers declined in the post-1920 period (Table IV).

The largest immigrant groups to enter the industry in the north Pacific region were English, German and Scandinavian (see Table V). Because the Industrial Revolution came earlier to Great Britain, English immigrants drew on a century of working class tradition. By 1900 the English "hereditary proletariat" was already molded into trade unions and by 1906, the British Labour Party had been formed to seek reforms which would lessen the burdens on the working class.[19] Thus, English immigrants to British Columbia arrived at a time when their experience could significantly aid the budding industrial union movement in the province's woods and mills. The IWA's present-day integration with British Columbia's New Democratic Party and its almost universal adherence to industrial union principles can undoubtedly be traced to these English working class immigrants.[20]

Compared to the English, the German working class developed late. When the English Luddites were trying to stop the technological displacement and attendant misery that capitalism brought to the previously pastoral English countryside, German workers were still tied to the land and to feudal traditions.[21] The heaviest German immigration to the Northwest came in the middle and late nineteenth century. Since many of these immigrants came from agrarian and recently feudal backgrounds, they were unlikely to have any experience in industrial unionism. In addition, most of them only worked part-time in small mills and farmed the cutover timberland to supplement their mill incomes. These "stump ranchers," while numerically significant in the Pacific Northwest and in British Columbia, never expressed themselves in an ethnically identifiable bloc within the industry's unions. The few German socialists who might have been part of this early migration to the region found that a centralized industrial work force did not exist, because at the time of their arrival, the wood products industry, especially in British Columbia, was still in its infancy.[22]

The third major immigrant group, the Scandinavians, came in two waves. Like the Germans, those immigrating before 1890 were predominantly agrarian and not experienced in industrial unionism. Those who arrived later, however, were mostly from industrial origins[23] where they had been influenced by Marxist theory so that they "displayed a corresponding aversion to the capitalistic form of society." Like capitalism everywhere in the late nineteenth century, Swedish capitalism was seeking expansion

TABLE III.

**Percentage of population foreign born,
Oregon, Washington, and British Columbia, 1850 - 1970**

	1970	1960	1950	1940	1930	1920	1910	1900	1890	1880	1870	1860	1850
Oregon	3.2	4.0	5.6	8.1	11.1	13.7	16.8	15.9	18.0	17.5	12.8	9.8	7.7
Washington	4.6	6.3	8.2	11.5	15.6	19.6	22.4	21.5	25.2	21.0	21.0	27.1	—
British Columbia*	22.7	25.4	29.0	37.0	46.0	53.3	—	44.2	42.1	29.2	—	—	—

* The Canadian census counts only *non-British* persons born outside Canada as "foreign born." The figures in this table have been corrected to include all persons born outside Canada as foreign born.

Sources: U.S. Department of Commerce, *Bureau of the Census, Census of Population: 1970* (Washington: U.S. Government Printing Office), Vol. I, Parts 39 and 49, Table 45; *1950*, Vol. II, Parts 37 and 47, Table 24; *1940*, vol. II, Parts 5 and 7; *1930*, Vol. III, Part II, Table 13; *1920*, Vol. III, pp. 512, 989. Statistics Canada, *1971 Census of Canada* (Ottawa: Minister of Industry, Trade and Commerce), Vol. I, Part 3, Table 126; *1950*, Vol. I, Table 45; *1940*, Vol. III, Table 19, *1930*, Vol. IV, Table I. Dominion Bureau of Statistics, *Canada Year Book 1922-23* (Ottawa: Minister of Trade and Commerce), p. 167.

TABLE IV.   **Laborers in Oregon and Washington mills,
by birthplace, 1910 and 1930**

	1910			1930		
	Total	Foreign Born	Percent Foreign Born	Total	Foreign Born	Percent Foreign Born
Oregon	7,629	3,058	41	11,926	2,326	19
Washington	16,908	7,752	47	19,127	5,567	30

Sources: U.S. Department of Commerce, *Bureau of the Census, Population 1930,* Vol. VI (Washington: U.S. Government Printing Office), pp. 1367, 1707; 1910, Vol. IV, pp. 507, 527.

through imperialism and, therefore, demanded compulsory military service from its working class. This factor, coupled with the general social instability inherent in capitalism, encouraged the emigration of class-conscious Swedish workers.[24] Like their English counterparts, this group of Scandinavians was able to play a prominent role in unionizing the North American timber industry.

Of all the ethnic groups which labored in the wood products industry and joined the IWA, the Scandinavians have, perhaps, been the most visible. With the notable exception of the International's first president, Harold Pritchett, who was born in England, many of the union's best organizers — Ernie Dalskog and Hjalmar Bergren among them — came from Scandinavian families. Those who participated in the 1930s drive to organize loggers in British Columbia were "almost all Scandinavians. . . . The Canadian, the native Canadian in the B.C. woods, did *damn* little to organize the trade unions in the earlier years."[25] In Washington, Ralph Nelson and Karly Larsen, Swedish and Danish respectively, were prominent among those who built the union and fought to preserve its progressive character. In the midwest, it was the Finnish names — Matt Savola, Ilmar Kouvinen — which stood out.

These immigration patterns significantly influenced later developments within the IWA. First, the early settlement of German farmers in the lower

TABLE V   Foreign born in Oregon and Washington, by selected countries of birth, 1850-1979 (in thousands)

	1970	1960	1950	1940	1930	1920	1910	1900	1890	1880	1870	1860	1850
Oregon													
Foreign Born, Total	66.1	71.3	83.6	87.6	106.7	102.1	103.0	65.7	57.3	30.5	11.6	5.1	1.0
England/Wales	6.4*	5.3	6.4	6.7	8.6	8.5	8.5	6.0	6.0	3.1	1.4	0.7	0.2
Scandinavian													
Norway	2.5	3.9	5.3	6.1	7.4	6.9	6.8	2.7	2.2	0.5	–	–	–
Sweden	2.3	4.5	6.9	8.4	11.0	10.5	10.0	4.5	3.7	0.9	0.2	–	–
Germany	6.6	6.9	7.9	9.8	12.9	13.7	17.7	13.2	12.4	5.0	1.8	1.0	1.0
Finland	N/A	2.2	3.5	4.3	5.5	6.0	4.7	2.1	–	–	–	–	–
Washington													
Foreign Born, Total	25.6	178.6	191.0	203.1	244.6	250.0	241.1	111.3	90.0	15.8	5.0	3.1	–
England/Wales	2.3*	14.0	15.8	17.2	21.9	22.8	21.4	12.0	11.4	1.8	0.8	–	–
Scandinavian													
Norway	1.8	18.5	23.3	26.4	31.4	30.3	28.3	9.8	8.3	0.5	0.1	–	–
Sweden	1.2	13.5	20.9	26.9	34.0	34.7	32.1	12.7	10.2	0.6	0.1	–	–
Denmark	0.3	3.3	4.7	5.7	7.1	8.3	7.8	3.6	2.8	0.2	–	–	–
Germany	2.3	14.4	12.9	15.4	20.5	22.3	28.8	16.6	15.3	2.1	0.6	0.5	–
Finland	N/A	4.7	7.2	9.1	11.0	11.8	8.7	2.7	–	–	–	–	–

* United Kingdom

Sources: U.S. Department of Commerce, *Bureau of the Census, Population 1970* (Washington: U.S. Government Printing Office), Vol. I, Parts 39 and 49, Table 49, *1940*, Vol. I, *1940*, Table 99, *1940*, Vol. II, Parts 5 and 7, Table15; *1950*, Vol II, Parts 37 and 47, Table 24.

Columbia River and Willamette River valleys was the foundation upon which the labor movement in the Portland area was built. These were the early workers who had come looking for farming opportunities, bringing with them little industrial or socialist experience; therefore, work in the woods or mills was incidental to their primary goal of self-sufficiency. The mills which they worked in were small in comparison to those which were built later on Puget Sound or in British Columbia because there was no large concentration of capital such as that which marked the late nineteenth century. Without large-scale industrialization, the conditions for radicalism and industrial unionism were absent.

Second, immigration came later to northern Washington and British Columbia, which meant that the newcomers brought with them a more advanced political attitude from their experience with industrial unions and socialist movements in England and the Scandinavian countries. When they arrived, concentration of production into larger units had already taken place. Thus, the social conditions were more favorable for industrial unionism and labor radicalism.

Immigration was, of course, only one factor that was affected by the uneven regional development of the industry. The evolution of the North American labor movement also contributed to the emerging political imbalance that would tear the IWA apart soon after its formation.

Working Class Organizational Development: Craft Unionism in Oregon, Industrial Unionism in Washington and British Columbia

The Portland, Oregon, labor movement developed an early reputation for conservatism. One historian has noted that in the 1880s and 1890s "...nowhere else in the North Pacific Region were labor leaders such models of restraint and caution as in Portland." The AFL was introduced to the Pacific Northwest through Portland labor leaders in 1887, and for several years Portland stood alone in the vanguard of Samuel Gomper's "pure and simple unionism."[26]

A more radical union tradition was established in Washington and British Columbia. The coal mining communities of northern Washington and Vancouver Island attracted the attention of the Knights of Labor who gave organized expression to the anti-Chinese movement and attracted hundreds of workers who felt threatened by the Chinese laborers brought in for railroad construction work. The left wing of the Knights, however, shunned the racial politics of the mainstream and linked up with the International Workingmen's Association which claimed to be affiliated with the Marxist Red International.[27] It was this left-wing legacy that guided radical woodworkers in northern Washington and British Columbia in their early attempts to organize.

The lumber industry often followed in the footsteps of mining and railroads. The need for timbers to shore mine shafts, and the demand for railroad ties and lumber for general construction along new rail lines provided a profitable stimulus to found new sawmills. Similarly, union organizing among miners often preceded and influenced woodworkers. The area around Coeur d'Alene, Idaho, and the Kootenay region of southeastern British Columbia — areas destined for bloody labor strife in the lumber industry — were also the scene of pitched battles fought by the Western Federation of Miners. In the South, the IWA later took over entire locals of sawmill workers which had been organized by the United Mine Workers.

Prior to 1905 there had been few attempts by woodworker organizers to tap the radical heritage of the Pacific Northwest and most of these attempts had been failures. In 1890 an unsuccessful strike during a period of depressed economic conditions broke an attempt to organize shingle weavers; it was not until 1903 that the International Shingle Weavers Union of America was formed in Everett, Washington. A 1905 strike by Shingle Weavers in Ballard, Washington, was extended to 365 mills in the area but the strike was lost when the mill owners organized the Shingle Mills Bureau to fight the union. During 1905–06 the International Brotherhood of Woodsmen and Sawmill Workers was organized and received a charter from the AFL. The Royal Loggers was also formed in 1906, but both unions failed because of the irregularity of employment, the scattered locations, and the transience of the workers.

These early efforts to organize also suffered from a lack of support from the nationally-based AFL, which was only accustomed to organizing skilled workers into homogeneous unions along craft lines. Thus, carpenters, plumbers, loggers and sawyers would have all been in separate unions even though they worked together in the same industry. Unskilled workers in mass production industries such as lumber did not fit into any of the AFL's neat craft categories, and consequently they were often neglected.

The AFL approached lumber workers skeptically because they were mostly unskilled, often foreign-born, and prone to radicalism. The Brotherhood of Carpenters, into whose jurisdiction the woodworkers would have fallen, were interested in having jurisdictional hegemony over them, but not in organizing them because they feared that these "uncarpenterlike carpenters" would swamp the craft-oriented structure and "attempt to foist industrial unionism upon the carpenters."[28] The organization of the "timber beasts," as early woodworkers were called, had therefore to await the formation of the Industrial Workers of the World (IWW).

The IWW

The Industrial Workers of the World (IWW or Wobblies) was born in 1905. Its purpose was to organize all workers regardless of race, color, creed, national origin, or skill into "One Big Union." Wobblies avoided working within the political system, arguing that all political institutions were "bourgeois" and therefore inappropriate for workers to use. They also frowned on most standard organization techniques; they were against established leaderships, dues, and contracts with employers. They stressed the use of the strike weapon and the building of direct worker control over the means of production. They were famous for building (though not always successfully) community support for their actions by involving everyone in their general strikes and cultural activities.

Their tactics and strategy were appropriate to the industry at that time since the sledge-hammer tactics of the boss lumbermen warranted no greater sophistication. The operators' use of vigilantes and violence against the IWW and their reluctance to provide civilized working and living conditions could only be redressed on the job and in the streets.

The west coast timber workers who had become very skeptical of elitist craft-unionism and rampant class collaboration in the AFL received the new union warmly.[29] Transient workers were especially attracted to the IWW. While some of them were foreign-born and inclined toward socialist politics, the others were just as likely to be native-born and imbued with a deep sense of personal independence and adventure. Moving as often as they did, these "hobo" workers were often unable or ineligible to vote and could never benefit from a contract signed in any one camp. Thus, they scorned electoral politics and collective bargaining and became the catalysts of a sizable syndicalist bloc in the IWW. In 1908 it was this syndicalist "Overalls Brigade" from Portland, Oregon, that drove out the more politically-oriented eastern intellectual bloc, led by Socialist Labor Party founder Daniel DeLeon. In 1913, these Western delegates again provoked divisions when they tried to abolish the general executive board of the IWW and give the affiliated unions complete autonomy.[30].

The first organized IWW activity in the Northwest was a 1907 sawmill strike in Portland, Oregon. The AFL, through the Central Labor Council of Portland, denounced the strike and the IWW's role in it, and as a result the strike was lost. However, it established the Wobblies[31] as a working class force and for more than a decade, neither the AFL nor the business community would be able to ignore them.[32]

Despite defeats similar to the one in Portland and the very brutal Everett Massacre in 1915, the Wobblies made gains for timber workers through the

use of direct action tactics. Tom Scribner, an IWW member, later recalled
the struggle for improved living conditions in logging camps.

> When we launched a campaign to make the employers furnish bedding in the
> camps...we had five-man committees to meet the trains, boats and busses. The
> five-man committees would ask the workers to give up their blankets and they
> would explain why....The blankets would be heaped up in large piles and
> burned on the streets. It wasn't very long until the signs at the employment
> offices asking for "one choker setter, one bucker, must have blankets" com-
> pletely disappeared simply because there were no more men with their own
> blankets.[33]

Direct action tactics like these also won the eight-hour day and the end of
double-decker bunks in logging camps.

The IWW convened in Spokane, March 5 and 6, 1917, to form the Lumber
Workers Industrial Union, and in July the IWW and AFL unions struck for
the eight-hour day. The employers retaliated by forming the Lumberman's
Protective Association to fight the union's demand. Immediately the cry was
raised by the employers and the government that the strike was hurting
spruce production essential to the manufacture of airplanes for the war, but
despite the urgings of Governor Lister of Washington, the employers refused to end the strike by granting the eight-hour day. By September the
Wobblies had moved their fight to a "strike on the job" in order to prevent
scabbing and raise finances. "Instead of doing a day's work they practised
'conscientious withdrawal of efficiency.'"[34]

The government responded by sending Lieutenant Colonel Brice Disque
to the Northwest. Disque, a self-styled Progressive reformer, had been influenced by Samuel Gompers. Sympathetic to the needs of the War Department, Disque collaborated with the boss lumbermen to stabilize the lumber
industry, and when he was finished, "neither the AFL nor the IWW threatened
employers' economic power." The workers got their eight-hour day and
improved working conditions, but in return Disque "closed the woods to
labor organizers and to trade-union members by organizing a company
union — the Loyal Legion of Loggers and Lumbermen...with practically
compulsory membership and a no-strike policy."[35] In addition, Disque har-
nessed soldiers in civilian clothes to work in the woods for civilian wages.
This "Spruce Production Division" of the U.S. Army operated under military regulations and added the cover of patriotism and military authority to
the 4-L's nearly insurmountable power. At the end of 1917 the 4-L had
10,000 members in 300 locals; by April 1918, there were 70,000 members.
The 4-L continued after the war as a pure and simple company union, but
by 1922 its membership had dropped to 10,000 and there it hovered for
several years.[36]

The last chapter of the IWW was essentially written with the post-World War I "red scare" that began when U.S. Attorney General A. Mitchell Palmer embarked on a campaign to arrest, prosecute and deport radical agitators. The pretext for the campaign was the fear that Russian Bolshevism would spread unless positive steps were taken to prevent it. Palmer established what was to become the Federal Bureau of Investigation (FBI) and carried out much of his work in close co-operation with employers and with the American Legion, which had just been formed at the end of the war. Palmer's assistant was J. Edgar Hoover, who was to become famous in later years as the head of the FBI.

The campaign began in 1919 and reached its high point on the night of January 1, 1920 when "ten thousand American workers, both aliens and citizens, most of them trade union members and many of them union officials, were hauled from their beds, dragged out of meetings, grabbed on the street and in their homes, and thrown into prison."[37] The IWW was the main target of the raids.

The main attack on the IWW in the northwest had come in Centralia, Washington, on Armistice Day, 1919, when a parade of American Legionnaires halted in front of the IWW hall and then charged it. The Wobblies defended themselves, and when the gunfire was over, three Legionnaires lay dead. Wobbly Wesley Everest was chased to the Skookumchuck River where he shot and killed one of his pursuers before he himself was captured. That night, Everest was taken from the jail, castrated, shot and hung. The coroner cynically reported to the Elks Club that "Wesley Everest had broken out of jail, gone to the Chehalis River bridge and jumped off with a rope around his neck. Finding the rope too short he climbed back and fastened on a longer one, jumped off again, broke his neck and then shot himself full of holes."[38]

Eleven Wobblies were charged with the murder of the Legionnaires. When their defence tried to show that a grand conspiracy existed on the part of the State of Washington to rid itself of the IWW, the court disallowed their evidence. The jury found two of the defendants guilty of "only" third-degree murder, but the judge would not accept the verdict. A second verdict, more to the judge's liking, found seven guilty and one insane.

Subsequent investigation has shown that the trial was a perversion of justice. Most of the jurors eventually recanted their decisions in sworn affidavits; it seems that the Red Scare had temporarily emboldened and encouraged them to send seven innocent men to prison. Some of the Wobblies died in prison, the others were eventually released. The last of them, Ray Becker, was freed in 1940 after the IWA and his friends had formed a "Free Ray Becker Committee" in his behalf.[39]

The Centralia case contributed to the anti-Communist hysteria and "witch-hunts" that kept the Northwest woods unorganized throughout the 1920s. The Russian Revolution, moreover, produced new tensions within the IWW when the Communist Party with its prestigious associations with the Revolution began to attract IWW members. In 1921, the IWW was invited to join the Communist Red International of Labor Unions, or the Profintern; debate over affiliation caused violent divisions within the IWW, with the main opposition coming from lumber workers in the Northwest who had established their reputation as libertarian free spirits opposed to all capitalist, trade union, or socialist forms of organization. Expelled for berating the IWW executive board for its Communist sympathies, the Portland "Emergency Program" established its own IWW branch led by James Rowan.[40]

The tradition of anarcho-syndicalism which was so appealing to the migrant lumber workers thereafter relegated them to a marginal role in the organizing boom of the early 1930s, and at a critical point in the struggle between the AFL and the emerging IWA-CIO, the IWW would remain "neutral." By the end of the 1930s, the Wobblies' anti-Communist leaders who were embraced by the CIO's national office in its campaign against the Communists would still proclaim their allegiance to IWW principles. But former IWWS, many of whom were Scandinavian settlers in Washington or British Columbia, would emerge as leaders of the Communist Party's National Lumber Workers Union.

The Centralia conspiracy was a benchmark in the development of class relations in the wood products industry. The heyday of timber robber-barons was nearly over. The great woods of North America had been cut over from east to west, there were no more waves of immigrants to exploit and the success of the Bolshevik Revolution had heralded the rise of working class consciousness everywhere — even among North American woodworkers. Confronted by these new realities, timber operators had dug in for a fight and summoned the state for help. The intervention of the state had left the Wobblies defenceless, because it was almost impossible to counter state power with the IWW's limited "point of production strategies" for revolution. Thus, the IWW had been swept aside by the epitome of governmental unionism — the 4-L.

By the outset of the 1930s, then, the historically uneven social conditions resulting from the development of the wood products industry had produced a politically fragmented work force. The Willamette Valley running south from Portland and the lower reaches of the Columbia River would continue to reproduce conservative business unionism as they had since the founding of the AFL, while the Scandinavian workers who arrived later in northern Washington and British Columbia to work in the large mills would provide an element of radicalism. Finally, the materially different circumstances of immigrants, native born migrants and the "home guard" workers who were indebted to the company produced divisions that resulted in erstwhile fellow workers gravitating to opposing camps by the end of the 1930s.

2

Community and Union Organizing in the Depression Years

The 1920s had been lean years for union organizing in the woods and mills. Except for the government-sponsored Loyal Legion of Loggers and Lumbermen, only the Communists had persisted in their attempt to organize woodworkers. And when depression conditions threw thousands out of work, the Communists had maintained their reputation as committed organizers by working in the unemployed and relief movements in mill towns and logging camps. The strong bond thus developed between Communist Party activists and working class communities enabled leftists to assume union leadership when efforts to form an industrial union were resumed in the mid-1930s.

By 1929 the lumber industry was plagued by chronic overproduction and competition from building material substitutes — concrete, brick, tile, composition shingles, fiberboard and hundreds of steel and metal commodities. Lumber consumption and prices fell; home building dwindled. Lumber companies fought back by attacking the prevailing wage structures. As a result, on the eve of the Great Depression, the average wage for lumber workers (except planing mill machine feeders) had dropped below the level for 1919.[1]

Depression unemployment struck hard at the ranks of sawmill workers and loggers, and no organization existed to combat the arbitrary layoffs, shortened work weeks and wage reductions. From 1929 to 1933, employment in the Oregon lumber industry dropped from 32,532 to 13,007, a decline of sixty percent. During the same period, total annual wages in the state's lumber industry plummeted from $51 million to $11 million, a drop of seventy-seven percent.[2] Building construction suffered a similar fate. Of the 2,022 construction workers employed in 1929 in Portland, the economic centre of Oregon, only 689 held jobs by 1933. The slowdown in construction and lumbering cut sharply into the ranks of sawmill workers. In 1929, 3,763 men had labored in Portland's sawmills; by 1933, 2,700 of the city's 17,000 unemployed were former sawmill workers.[3] Up and down the Pacific Coast and in the short log industry east of the Cascade Mountains, the faltering economy displaced loggers and sawmill workers, so that by 1932, less than twenty percent of the lumber industry in the Pacific Northwest was operating, and about half the men employed in it were working on a short-day basis.[4]

The minimum wage for lumber workers kept dropping. For example, in early 1930, the Clark and Wilson Lumber Company which cut virgin fir on the headwaters of the Nehalem River between Vernonia and Timber, Oregon, started hook tenders at $9.60 per day, but ten months later they were starting them at $4.60. Other loggers in their camps received 22 cents per hour, but they had to work seven days a week plus some overtime to avoid owing the company at the end of the month. Bed and board was $1.35 daily, but "...the company never cut that."[5] Finally, competition from mills and logging camps, some of them paying as little as $1.50 per day, forced even the company-oriented Loyal Legion of Loggers and Lumbermen to abandon its wage scale of $2.60 per day.[6]

Safety regulations and inspections were either non-existent or ineffectual, and every week loggers fell beside the trees. The companies wanted production and demanded the "speed-up". Bucking and falling teams from different camps were forced to compete by foremen who split a percentage of the day's cutting with management. They would fire the last man up the hill to the falling site in the morning or out at night, regardless of his speed, as a lesson for the other workers; there were always more young, unemployed loggers waiting. In order to learn the job-site location, workers had to pay the "job shark," the man who posted work openings where the loggers gathered. Collusion between the foreman and shark was not unusual; thus, the more men fired and new men hired, the more money for both foreman and shark.

Sawmills and shingle mills were also places of frequent injury or death. Fingers, hands, and limbs were easily severed by inadequately guarded mill machinery. Rough cut lumber, passing quickly along live rollers and conveyor belts, frequently jammed and caught the unwary. Pieces of lumber, kicked out by saws and edgers, became veritable missiles. Falls from the numerous mill platforms were not uncommon as workers hurried from task to task. The noise of the whirling saws, the gears and chain-drives of the machinery deadened the workers' senses and slowed their reflexes, while heat and sawdust contributed to high tubercular and asthma rates for mill and shingle workers.[7]

The deteriorating working conditions, high unemployment and low wages that accompanied the onslaught of the 1930s spawned walkouts and sporadic, quick strikes led by short-lived camp or mill committees. Those workers who had taken part in the struggles of the IWW began to circulate union leaflets and raise questions about working conditions and the future of the lumber industry.

The Communist Party and Lumber Workers

The first significant organizing during the early depression years was conducted by the Communist Party's Trade Union Unity League (TUUL) and its wood affiiliate, the National Lumber Workers Union (NLWU). The Communist Party had formed the Trade Union Education League in 1921 to allow party members to work within the existing labor unions and avoid the anti-Communist harassment that had been prevalent during the post-war years. But the conservative and anti-Communist AFL executive board and the leaders of the national AFL unions were powerful enough to effectively block Communist Party activities within the unions. Consequently, in 1929, this "boring from within" strategy was abandoned and replaced by the "dual unionism" of the Trade Union Unity League. Communist labor organizations now planned to develop industrial unions to compete with the craft-biased AFL.[8]

In September, 1929, the Trade Union Unity League organized the National Lumber Workers Union (NLWU) and established a local in Seattle. The League hoped to unite all forest and mill workers from tree stump to the finished product in "one big industrial union" similar to the old Wobbly vision.[9] The extensive demands made by the new union not only reveal in detail the conditions under which woods workers and mill employees were laboring ten years after the IWW's demise, but they also provide a yardstick against which to measure lumber workers' progress today.

> A minimum wage of $6.00 a day for all workers in the industry, with double time for overtime.
> A 5-day, 35-hour week with a full week's wage.
> A 7-hour day from camp to camp.
> Abolition of the Clearing House and the blacklist.
> The elimination of the Four L.
> The abolition of the "gypo" and piece-work systems.

Free employment service under workers' control.

Full pay for loss of time caused by shutdowns due to defective machinery, repairs and other such causes.

An annual vacation of not less than two weeks with full wages.

The elimination of the speed-up system.

Equal pay for equal work for men, women and young workers.

Abolition of child labor. The 6-hour day for young workers.

Installation of adequate washing and toilet facilities in all camps and mills.

Shower baths and drying rooms to be installed in all camps to be kept in good running order.

Elimination of unsanitary conditions in the camps and proper garbage disposal.

Wholesome food in the camps.

Well ventilated, heated and thoroughly clean houses for camp workers with one worker to a room.

Clean blankets and sheets.

Free access to all working class newspapers and literature at all times.

Installation of safety devices in the camps and mills to protect health and life.

Compensation laws providing full wages when a worker is injured by accident. Full medical, hospital and compensation expenses to be paid by the employers.

Social insurance against unemployment, sickness and old age, to be paid by the employers and managed by the workers.

Immediate release of the Centralia and other class war prisoners.[10]

The NLWU was at times a very loose organization; usually, only a handful of men were active in any one logging camp, and funds were scarce.[11] An extensive "blacklist" and the availability of unemployed, non-union workers hindered the union's organizing attempts. However, the union launched several strikes in Washington, at Moclips (1930), Anacortes (1931), and Willapa Bay (1933), successfully stopping or reducing proposed wage cuts.[12] In Oregon, union members supported a few small strikes and marched in Portland's unemployment demonstrations, but did not make a serious effort to organize the city's sawmill workers.

After four years of activity, the National Lumber Workers Union convened for their first national convention in Seattle on December 18, 1933. Thirty-nine delegates attended, representing 2,800 members who worked principally in the Pacific Northwest. Union president James Murphy, a former IWW member and a Communist, called for higher wages, better working conditions and unemployment insurance.[13]

In British Columbia the equivalents to the Trade Union Unity League and the National Lumber Workers Union were the Workers Unity League and the Lumber Workers Industrial Union (LWIU).[14] All of them were part of the U.S. and Canadian Communist Parties' policies to develop effective national unions for industrial workers outside the structure of the AFL or of the Trades and Labor Congress (TLC), which was the AFL counterpart north of the border.

The LWIU of Canada was far more successful than the Northwest's NLWU. Its roots stretched back to the OBU–the One Big Union of the early 1920s. The birth of the Lumber Workers Industrial Union under the Workers Unity League in 1929 gave Communist lumber workers in British Columbia another opportunity to learn and practice the skills which would serve them well in the drive for industrial unionism in British Columbia and the Pacific Northwest.

Strategy and Tactics under Depression Conditions

The first significant opportunity for the LWIU to implement its organizing campaign occurred at the Canadian Western Lumber Company's mill at Coquitlam, fifteen miles up the Fraser River. There, the company had built the largest mill in the British Empire and the second largest in the world; the sawmill, a plywood plant, a fifteen-machine shingle mill, a planer mill, and dock facilities for world-wide shipping stood on two hundred acres of former Indian Reserve land.[15]

Since the founding of the mill shortly after the turn of the century, the operation had been stoked by needy French-Canadians shepherded to British Columbia by a Catholic priest named Father Maillard. The company financed both their travel west and the construction of a townsite named "Maillardville" outside the mill gates to house the new arrivals. Inside the gates, Japanese, Chinese and East Indians lived in a shanty-town called Fraser Mills, controlled entirely by the company. Scrip was available to all 1,800 workers. Men with large families made regular use of the company denominations, so that at month's end their pay only cancelled their scrip indebtedness to the "company store."

When the depression came, the management at Fraser Mills moved swiftly to counter its effects. In June, and again in July, and in August, 1931, they cut hourly pay rates by ten cents. They reduced the sixty-hour work week to six nine-hour shifts. New machines and renovations were brought in to increase production and speed up both the mill and the workers. Drivers and attendants for twenty-two of the company's horse-drawn lumber carriages were fired.

Even before the Depression had hit, progressive Fraser Mill employees had formed the Maillardville Ratepayers Association to monitor rent and electric bills; they had then merged with the Ratepayers Association in Coquitlam, the larger community which surrounded the millworkers' district. This progressive coalition had placed Scots immigrant and

Communist Tom Douglas on the Coquitlam municipal council in 1930, face to face with a conservative mayor and two lumber company executives. But the Ratepayers Association was impotent when confronted with Fraser Mills' pay and work reductions. The mill workers (some of whom were already secret members of the LWIU) were therefore willing to listen when Arnie Johnson, a Communist organizer for the LWIU, appealed for their membership.[16]

By September, 1931, not a wheel was turning at Fraser Mills. The union's demands for a ten-cent hourly increase, union committee recognition, abolition of the scrip system and renovation of the living quarters assigned to the Asian workmen had been overwhelmingly adopted by the workers, and just as convincingly rejected by Henry Mackin, president of the Canadian Western Lumber Company.[17]

For a new union, the organization and tactics used were remarkably sound. A strike committee of eighteen men represented all mill departments and nationalities, and strike support was increased with the formation of a ladies auxiliary. A twenty-four hour picket line with standby reinforcements closed the mill tight. Asian workers dug tons of potatoes which had been donated along with truckloads of vegetables by sympathetic Fraser Valley farmers. Single workers ate at the union restaurant, while married men obtained rations at an LWIU general store with the result that some French-Canadians with large families ate better on strike relief than they had on company wages. When the company persuaded the local priest to withhold absolution until the French-Canadians broke ranks and scabbed, LWIU organizers lined up their Model Ts in Maillardville and transported the workers to another church for services.[18] Sports activities, dances, and even free haircuts at an LWIU barbershop heightened enthusiasm for the three-month strike.

The chairman of the Fraser Mills strike committee was twenty-seven-year-old Harold Pritchett, who had worked in the mill since 1927. In 1912, when he was eight, his family had emigrated from Britain. Following graduation from grade school seven years later, Pritchett had begun work in a sawmill. On his first job he had tended conveyors in a planer mill for ten cents an hour, eleven hours a day. Within a year, he had learned the skills of sawing and packing cedar shingles, and in 1925, he joined the Vancouver local of the AFL Shingle Weavers Local.

By this time, as a result of his previous five years' work, his discussions with his workmates and regular reading of Marxist literature, he had developed into a politically astute union activist. He had come to believe that union action should be premised on the understanding that the interests of employers and employees were not the same. This put him in conflict with Percy Bengough, head of the Shingle Weavers Union and the Vancouver and New Westminster Trades and Labour Council. When Pritchett was elected head of the Shingle Weavers Local four years later, Bengough and the local mill operators persuaded William Green, president of the American Federation of Labor with which the local was affiliated, to intervene with the result that Pritchett was expelled from office. At about the same time, Pritchett joined the socialist Independent Labour Party, but he was soon recruited to the Communist Party by LWIU organizer Arnie Johnson.[19]

But the united action of lumber workers and organizers like Pritchett could not budge the management of the Canadian Western Lumber Company. By early November relief funds were desperately low. The union raised some money with boxing exhibitions given by pugilist loggers who donated their skills, and by charging admission to union dances, but enthusiasm and charity could not pay the mounting bills. Weekly rations for the 193 strike families were cut to:

½ lb. butter	salt
½ lb. sugar	vegetables
½ lb. lard	meat, fish or eggs, if we have it
½ tin milk	tea or coffee
3 lbs. flour	juice and other[20]

On November 20, after nearly three months, the strike came to an end. Workers voted overwhelmingly to return to their jobs under a new agreement: a ten-cent hourly wage increase, company recognition of the employees' grievance and safety committee, and non-discrimination against strikers. While the contract fell short of meeting the original demands, it was a real victory in the early 1930s when most strikes resulted in broken unions and further wage reductions.

The impact of the strike rippled far beyond Fraser Mills because the strikers' communicative process had been evolving. What had begun as small and amateurish memos, by November 1931 had become a four-page, five-cent newspaper, the *B.C. Lumber Worker*, the western organ of the Lumber (and Agricultural) Workers Industrial Union of Canada. The headlines in the first issue left no doubt which side it was on: "Wage Cuts in Canada — Wage Increases in Soviet Union;" "It is Defferent [sic] in Soviet Union;" "Defend the Soviet Union. Fight Against The Wage Cuts."[21]

The Fraser Mills strike also established precedents for minority relations. While the union failed to eradicate the company shanty-town for Asians, it nevertheless recognized the contributions that Hindu, Japanese and Chinese lumber workers made to the strike. If the workers had not challenged the employer's myth that Asians willingly accepted worse conditions and longer hours than whites, the Asians' fifteen percent of the lumber industry workforce would have remained unorganized and racism would have been rampant.[22]

1931 to 1934 in B.C.

The Lumber Workers' Industrial Union met with mixed success following the Fraser Mills strike. But partial victories during 1931 and 1932 at Elk River Timber, Campbell River Timber, Morgan's Camp, the Queen Charlotte Islands, and Camp 3 at Sproat Lake helped the LWIU build a reputation among woods workers which the CIO would capitalize on in 1937.

Throughout these years the LWIU honed its propaganda skills. The *B.C. Lumber Worker* operated like a grapevine, arousing workers and facilitating communication between camps. Workers' letters, which frequently began with "Dear Comrade," warned of camp dangers, low wages, or particularly reprehensible bosses. In order to escape the blacklist, they signed their letters "Anti-High Baller," "A Lumber Worker" or simply with initials or a union card number.[23]

Because bundles of the *Lumber Worker* discovered by company owners or camp foremen quickly became fuel for the office stoves, union organizers regularly played a journalistic game of hide-and-seek. To deceive these arbiters of union news, the organizers placed the papers inside parcels of clothes, boxes of chocolate, or disguised them with gift wrap and feminine handwriting. Once inside the camps, the newspapers were distributed in the bunkhouses or placed beside the toilets in the large outdoor lavatories.[24]

In 1934 the years of clandestine nighttime bunkhouse meetings and the secret entrance of organizers and newspapers into the camps culminated in the last big British Columbia lumber strike prior to the establishment of the CIO. The LWIU struck Bloedel Lumber Company in the spring of that year over wages and camp conditions. The union wanted a fifteen percent wage increase to a $3.20 daily minimum, union recognition and no work on Sunday.[25]

In April the company flew scabs into their camp at Great Central Lake. In response, 350 loggers trekked fifty miles through the woods from Parksville to picket. Townspeople along the march and wives of loggers supplemented their food supplies. By the second day the group had swollen to five hundred, while rallies in Vancouver raised funds and the mayor went on the radio to defend the strikers. For seventeen days they camped at Great Central and harassed the scabs so that it was impossible to move logs from the camp, although work continued inside the camp where the scabs were protected by a detachment of provincial police who were reinforced by a machine gun.[26] By the end, strike headquarters had been moved to Vancouver and upwards of 2,500 lumber workers had joined the massive strike.

During previous strikes, loggers had headed for the bright lights of Vancouver where they had quickly lost their meager savings, thus hastening their return to work before the strike was effectively settled. This time, however, a picket camp was established at a tourist campground on the Campbell River. A small tent city was erected to minister to the strikers; food was donated, sanitary conditions maintained and drinking forbidden. Loggers with armbands patrolled nearby taverns to keep striking workers sober and out of trouble.[27]

The union formed "flying picket squads" of fifteen to twenty men who would sneak through the woods at night to the back of a logging camp, while decoys distracted the police at the camp's main gates. Once inside the bunkhouse, the pickets would distribute literature and present the union's position on the strike. To augment the flying squads, a seine boat was chartered so that organizers could take the campaign to isolated coastal logging camps.[28]

After three months, public pressure forced the provincial government's Industrial Relations Board to establish a forty cents hourly ($3.20 daily) minimum wage law for the entire lumber industry.[29] Union officials hoped the workers would stay out until union recognition had been granted, but after three months without paychecks, most loggers were eager to go back to the woods immediately. Nevertheless, 3,000 B.C. lumber workers had been brought into the LWIU by the end of the strike. More important, the union had bounced back from its partial defeat at Fraser Mills in 1931 to rekindle the lumber workers' drive for industrial unionism.[30]

Unemployed Organizing

When workers were laid off or on strike, union activists often regrouped and joined their former workmates in Unemployed Councils organized by the Communist Party. These councils were organized in Seattle and Portland as well as in smaller towns, such as Sedro Woolley, Washington. They agitated for increased unemployment relief and organized soup kitchens and bread lines for the distribution of food.[31]

Demonstrations in Portland were typical of those held throughout the Pacific Northwest. On February 26, 1930, three hundred demonstrators marched on city hall carrying placards reading: "Work or Wages," "Defense of the Soviet Union," "No Eviction for Non-Payment of Rent," and "Seven

Hour Day and Five Day Week."³² On February 10, 1931, six hundred unemployed workers gathered outside city hall for an hour-long meeting. Fred Walker, Northwest Organizer for the Young Communist League and spokesman for the group, presented Mayor George Baker with a list of demands:

> Free streetcar rides for the unemployed and their families. The use of city property and vacant buildings for housing the unemployed who could not afford to pay rent, and a special tax on property in excess of $2,500.³³

The Unemployed Citizens League also attempted to supplement the inadequate efforts of the county and city "make-work" programs. The League had been organized in Seattle, Washington in 1931 by Carl Branin of the Seattle Labor College, and within a year it had attracted 50,000 supporters. They were a curious mixture of political activists, ranging from former Wobblies and Communists to middle-class homeowners who wanted to maintain their property and keep food on the table. The League's political program included unemployment relief, free medicine and dental service, a thirty hour week for public employees and free burial for families in need. It was also a "self-help" agency which organized unemployed men and women to share their services and products. Within a year it had enrolled 6,237 members in twenty-six locals around Portland, Oregon,³⁴ and by the summer of 1932 it had spread to several other cities as well.

Organizing the unemployed was essentially the bailiwick of the left since many of the jobless were loggers and semi-skilled mill workers who were of no immediate concern to the Northwest's AFL craft union affiliates. Characteristically, the Portland AFL's Central Labor Council found the Communist leadership of the Unemployed Councils and their demonstrations more unsettling than the plight of the unemployed, and therefore usually refused to support unemployment demonstrations led by left-wing unemployment organizations. They also refused to allow representatives of the Unemployed Councils to address the AFL. By late 1933 the clarion call of "Keep Your Faith in America, Don't Let America Down" was going out to the unemployed through the pages of the AFL's *Oregon Labor Press*.³⁵ To the pundits of the AFL, the regular and militant unemployment demonstrations raised the spectre of Communism:

> If America should go Red? What would you do...? Now is the time to take some serious thought of this matter and see to it that there will be no regrets when you look back a year from now upon the years 1933–1934 and its winter of discontent.³⁶

Similarly, in British Columbia, unemployed lumber workers found most

of their support outside the craft unions. It was the organizations affiliated
with the Workers Unity League that led the struggle to organize the depression's unemployed. However, the price for leadership of unemployment demonstrations in British Columbia often matched that in the Pacific
Northwest: the police unlimbered their clubs, and arrested the demonstrators.

In spite of these events, the impact of radicalism on both employed and
unemployed lumber workers between 1929 and 1933 was slight. True, many
workers felt alienated from society and relief clothing and food lines further
accentuated the divisions between employed and unemployed, between blue
collar and white.[37] But the majority of lumber workers in the U.S. and
Canada did not participate in an organization of unemployed workers or an
unemployment demonstration.[38] And the National Lumber Workers
Union, the only militant industrial union in the Pacific Northwest, never
received unanimous support because of the scarcity of mill work and
logging jobs and the fact that the unemployed often accepted jobs in non-
union operations. Consequently, lumber workers failed to make really
significant progress toward industrial unionism until the 1934 Longshore
Strike had revealed the effectiveness of united militant action.

1934 Longshore Strike

Until 1934, a longshoreman's job security was tied to the paternalism of
the work-crew foreman; whisky, money and other assorted favors guaranteed jobs. Men milled about outside company offices at early morning
shape-ups until a signal from the boss indicated that they had been selected
for the day's stevedoring. Then, in May, 1934, the jurisdiction of the company unions which were maintained by the West Coast shippers was challenged by the International Longshoremen's Association, AFL. The Waterfront Employers' Association of San Francisco refused the ILA's demands for
union recognition, a dollar an hour wage, a thirty-hour week and a union-
controlled hiring hall.[39] Led by Harry Bridges in San Francisco, longshoremen retaliated by shutting down the waterfronts from San Diego to Seattle
and pressing for a coast-wide working agreement.

From May 9 to July 31, 1934, docks along the Pacific Northwest were
controlled by the striking longshoremen. Normal shipments of lumber and
agricultural products were curtailed, and sawmills were forced to close
when lumber could no longer be shipped by water. By June, with no end in
sight, 17,000 lumber workers were laid off and payrolls were slashed almost
in half. However, members of the National Lumber Workers Union solici-
ted funds for the longshoremen and spoke in favor of the strike; as a result,

lumber and sawmill workers supported the strike and did not scab despite their own desperate straits. Even the big mills at Longview, Washington ground to a halt when for four days, beginning June 19, 1934, sawmill workers there went out on strike in sympathy with the longshoremen.[40]

On July 3, San Francisco businessmen announced that their trucks and drivers would move through the picket lines to the piers along the Embarcadero and remove the goods stranded there since the strike began. Longshoremen attacked the trucks with bricks, and police retaliated with clubs, tear gas and gunfire. Two days later, on Bloody Thursday, the Battle of Rincon Hill left two longshoremen shot to death by the police, thirty wounded and forty-three more either clubbed, gassed or stoned. Four days later, 15,000 longshoremen and sympathizers marched up Market Street behind a flat-bed truck carrying the coffins of the slain longshoremen. Union sympathies were now cemented, and on July 16 a three-day general strike began in San Francisco.[41]

Anti-radical hysteria engendered by the general strike spread quickly. West Coast police departments sided unequivocally with the shippers and invested themselves with the kind of patriotic fervor reminiscent of the Palmer Raids of the early 1920s. On the first day of the San Francisco general strike, Portland police searched freight trains arriving from the south in an attempt to head off a feared invasion by "flying squadrons" of communist agitators; 130 men, mostly hoboes and migrant workers, were taken into custody. Two reputed labor agitators, who supposedly planned to "radicalize" the local longshore strike by promoting a general strike, were also found on the train.[42]

When local shipping companies demanded protection in order to continue loading ships along Portland's strike-bound waterfront, Oregon Governor Julius Meier ordered 1,000 National Guardsmen to mobilize immediately. Fortunately, the troops remained camped on the outskirts of the city after the Central Labor Council threatened to call an immediate general strike if the Guardsmen moved to the waterfront.[43]

Between July 16 and 21, 1934, even though the San Francisco general strike had ended and the ILA had agreed to arbitration, the Portland police continued their searches and seizures. Private residences, Communist Party headquarters, and the Marine Workers Industrial Union hall were raided. Union records and Communist literature were seized and taken to the police station. Three men were arrested for advocating criminal syndication, and thirty-two others were taken in for violations of the Oregon Criminal Syndication Act of 1919.[44] All of those arrested were closely associated with the Communist Party and had worked with the Unemployed Councils to keep the unemployed from crossing the longshoremen's picket lines.[45]

Dirk DeJonge, once the Communist Party's candidate for Mayor of Portland, was tried and sentenced on November 21, 1934 to seven years in
prison. The charges brought against him included advocating violence
during the longshore strike, being in possession of Communist Party literature, and conducting and attending Communist Party meetings.[46] After the
Oregon Supreme Court upheld the conviction and Dirk DeJonge had spent
nine months in the Oregon State penitentiary, the United States Supreme
Court, on January 4, 1937, unanimously decided that the lower courts had
erred in convicting him. The Court held "that the Oregon state law as
applied to the particular charge as defined by the state court is repugnant to
the due process clause of the Fourteenth Amendment".[47]

The results of the 1934 longshore strike did not pass unnoticed by loggers
and sawmill workers. The joint control of hiring halls with employers, the
thirty-hour work week, wage increases and exclusive bargaining rights won
by the longshoremen constituted notable union victories. More importantly,
however, the organization of the longshoremen meant that woodworkers
had a strong ally for their own union activities. The two groups of workers
were closely linked through family and occupational associations, while the
Communist Party tied together the activists in both unions. If the woodworkers struck, it seemed a virtual certainty that the longshoremen would
support them.

The 1935 Lumber Strike

In July, 1933, at Enumclaw, Washington, the AFL's Northwest Council of
Sawmill and Timber Workers was established with headquarters for the
Puget Sound District Council in Seattle, the Grays-Willapa Harbor District
Council headquartered in Aberdeen, and the Columbia River District
Council centered in Portland. These three bodies formed the basis for the
union's organizing efforts amidst the lumber workers.[48] This union was the
early beneficiary of Section 7a of Franklin Roosevelt's National Industrial
Recovery Act (NIRA), which guaranteed workers the right to bargain collec-
tively with their employers through representatives (unions) of their choice.

The NIRA also provided guidance and regulation for the industry itself,
which had been plagued for years by operators who had been overly zealous
in liquidating their assets — the trees — even though felling too much timber
frequently led directly to falling prices. As a remedy, the New Deal set up
the National Recovery Administration (NRA) Lumber Code which set wage
rates that varied from twenty-three cents per hour in the southern states to
forty-three and one-half cents an hour for northwest lumber workers. With
minor exceptions, eighteen was the minimum legal age set for employment,
and forty hours was set as the standard work week.[49]

In application, the NRA Lumber Code was a dismal failure. The lumber business was highly competitive, with thousands of camps and mills tucked away up river valleys beyond the reach of the NRA. The product prices, production requirements, and labor provisions of the Code were either not enforced or disregarded by operators. Numerous violations were allowed to go unchallenged by the NRA.[50] Even before the Supreme Court declared the NIRA unconstitutional in the Schechter Case (April 1935), the lumber industry had all but abandoned the Code, thereby becoming the nation's first major industry to junk Roosevelt's federal regulations.[51]

But although the Code collapsed, the NIRA's brief existence nevertheless served to clarify the positions of the unions on the left and the right of the political spectrum. Lumber workers were dissuaded from supporting the Communist Party's National Lumber Workers Union by the apparent "respectability" of the AFL's Sawmill and Timber Workers Union. The Loyal Legion of Loggers and Lumbermen (4-L), that faithful relic of the Wobbly-busting days of World War I, was recognized as the labor administrator of the Lumber Code by the employer's Western Pine Association and the West Coast's Lumbermen's Association, even though it represented less than ten percent of the Northwest's lumber workers. The 4-L changed its name to the Employees Industrial Union in an attempt to legitimize itself under the NIRA, but by the early 1940s it had passed from the scene.[52]

While employers did not take Section 7a of the NIRA seriously, ten thousand Northwest lumber workers showed that they did by demanding admittance to the Sawmill and Timber Workers Union. The AFL Executive Board was caught by surprise.[53] Since its 1914 convention, the AFL's United Brotherhood of Carpenters and Joiners (UBCJ), which claimed jurisdiction over lumber workers, had deliberately ignored them, because they did not want to admit unskilled workmen into the ranks of skilled craftsmen. But in the 1930s "...their din became too deafening to ignore,"[54] so the UBCJ established a so-called federal union to receive the lumber workers. Under this arrangement, the lumber workers' federal union was to be absorbed into the UBCJ as soon as collective bargaining was established. In the meantime, the AFL executive council retained the right to set policy and decide the ultimate fate of the hastily created union. By this strategy of chartering federal unions and tying them to the UBCJ, the AFL avoided any permanent solutions to the problem of mass industrial unionization.

At their March 1935 convention in Aberdeen, Washington, the Sawmill and Timber Workers Union was officially placed under the jurisdiction of the United Brotherhood of Carpenters and Joiners. By then, the Northwest's mills and logging camps were seething with discontent. During 1933 and 1934 lumber workers in that area had failed to receive any substantial wage or hour benefits by joining the union, and no major strikes had been called in the industry. Meanwhile, continued wage reductions and disillusionment with the benefits of the NIRA Lumber Code were stimulating strike demands.

A.W. "Abe" Muir, an executive board member of the United Brotherhood of Carpenters and Joiners, was sent west to assume control of the Sawmill and Timber Workers Union. Muir, who knew next to nothing about the rough-and-tumble world of logging and sawmilling, wanted to avoid a strike at all costs. The lumber workers, however, wanted better wages and conditions, no matter what the price, and almost immediately the union prepared demands for a new working agreement. The principal provisions of their proposed new contract called for union recognition, a six-hour day five-day week, overtime and holiday pay, seniority rights and a minimum wage of seventy-five cents an hour.[55] For the mid-1930s it was a significant proposal.

The employers insisted that they could not meet the new demands because of the unfavorable economic conditions in the lumber industry and they maneuvered to "wait out" the fledgling union. They claimed they were unable to meet even the NRA Lumber Code which had previously established a forty-two and a half cent per hour wage and a forty hour work week for the lumber industry. The workers, nevertheless, firmly believed that the industry could pay more, and they set May 6, 1935 as the day they would strike if the new contract had not been accepted by the employers. Weeks before the deadline, as negotiations began to break down, lumber workers began leaving the woods and mills. By May 9, the left wing paper, *Voice of Action*, estimated that 30,000 workers were on strike. An estimated fifty percent of the lumber workers employed in the Douglas fir region of the Pacific Northwest eventually stopped work.[56]

Although it was the largest the Northwest had ever witnessed, the strike was never totally solid. In general, the lines of support followed the Northwest's historical patterns of radical versus conservative unionism. The strike was weakest in conservative Portland where the locals were quick to back Muir's attempts to settle with the employers. The large Long-Bell and Weyerhaeuser local just down the river from Portland at Longview, Washington was one of the last to go out. However, left-wing support was also weakened when the IWW declared itself "neutral" after Communist leaders accused Muir of trying to sell out the strike; in the eyes of the IWW, the disciplined approach that the Communists took to union and political work was indistinguishable from the AFL's bureaucratic approach. The leaflets which the IWW issued to publicize its stand called the fight against Muir "an internal struggle within the corrupt AFL.[57]

At the local level, the strike brought to the surface the divergent interests of workers divided by skill level and political orientation, and it became the pivotal event that linked these divisions among lumber workers with the later factional struggles within the IWA. This continuity can be seen in the example of the Raymond, Washington local. Raymond is just south of Aberdeen on Grays-Willapa Harbor in southwest Washington in the transition zone between the lumber industry's conservative Oregon region and the radical regions in northern Washington and British Columbia. Thus, the area had been a crossroads for radical and conservative union tendencies for over twenty years. In 1911 and 1912, the IWW had led immigrant Greek, Austrian and Finnish workers in major strike actions against local lumber operators. But the owners loaded the foreign-born workers aboard ships and trains of 1912 and replaced them with men of family. By the end of World War I, the government-sponsored Loyal Legion of Loggers and Lumbermen (4-L) had established itself as firmly as the IWW had.[58] During the 1920s the complex social conditions of the Raymond area sustained the influences of both IWW radicalism and 4-L conservatism without causing friction in the community.

This situation changed after the AFL chartered a federal labor union local at Raymond on April 5, 1933. The workers had sought the local as a response to the Weyerhaeuser company's installation of a new management system at the plant. The system, called the Bedaux system, was named for Charles Bedaux, a controversial figure best known for his industrial systems of time and motion studies. The system installed at Weyerhaeuser was a productivity plan that "permitted more direct central management intervention in the direction, evaluation, and incentive of labor. . . ."[59]

The leadership of the new local was dominated by the semi-skilled planer graders whose control over working conditions was most threatened by the Bedaux plan. But when the workers gathered for a mass grievance meeting on September 5, 1934, the leadership declined to put a strike vote before the body. It opted instead for a petition campaign directed at the company's head office and at government agencies. According to historian Jeremy Egolf, "the conciliatory conclusion of the petition suggested the isolated and economically weak position of the Raymond mill workers, but also was indicative of the liberal social and political alliances of the [union's] leadership, which included ties with supervisory personnel for the purpose of fighting the Bedaux methods." This "less radical wing of the leadership," according to Egolf, "was composed primarily of semi-skilled workers and upwardly mobile straw bosses for whom Bedaux controls and speedup represented a regression to work patterns associated with a lower, less skilled status in the work hierarchy."[60] The leadership of the anti-Bedaux struggle remained predominantly in the hands of this conservative faction but they were opposed by a group that favored mobilizing the rank and file for strike action. The only successful actions against the new system were a few brief strikes led by this faction in the winter and spring of 1935.

In the spring of 1935, the anti-Bedaux struggle at Raymond was subsumed with other local struggles in the general lumber strike. The political trajectories of factions and individual leaders also became tributaries of the larger events. Some of the more conciliatory anti-Bedaux activists were promoted in the company or joined the emerging anti-Communist faction within the union. Some, like D.C. King, would become leaders of the anti-Communist bloc in the IWA. Others, like Clifford Kight, who had been active in the unemployed movement in the early 1930s and who later led a successful anti-Bedaux strike in South Bend, Washington, either joined or worked closely with the Communist Party.[61] Several of these individuals would be associated with the Communist bloc of the IWA a few years later. "Building a union in opposition to the Bedaux system enjoyed a broad base of support," concluded Egolf, but this solidarity broke down when the struggle came up against the limits to reform which established labor-management relations could yield. Those who were among the most militant about restoring control over their jobs became reactionary when they had to confront the broader ideological and political questions about the transformation of capitalism. Said Egolf:

> ...the expressed goals of the anti-Communist wing of the workers' leadership at Raymond were congruent with their individually ascendant positions within the local working class and their close personal and political ties with local small business and centrist elements in the Democratic machine. The liberals' ideas... were consonant with the...conservative elements' efforts to drive Communists and their allies from authoritative positions in the IWA during the late 1930s and 1940s.[62]

Two political traditions converged within the leadership of the anti-Bedaux forces. The priority placed on controlling the work process on the shop floor was an unmistakable legacy from IWW policy. However, the preference for a petition campaign rather than strike tactics was more in keeping with the social democratic traditions of the AFL and the 4-L. This merging of anarcho-syndicalist objectives and social democratic tactics was contradictory in most respects, yet both had shared a political base among the native born workers of this area. Moreover, both traditions carried with them a vehement disdain for communism and it was this common cause that bound them together.

In the Spring of 1935, however, the fight against communism was still nascent. More immediate was the dissatisfaction with Abe Muir's conduct of the strike. It was apparent to everyone that Muir was unwilling to lead a strike to uphold the original demands of the union. The more conservative strike leaders such as those in Raymond and Portland were willing to go along with his stand but the Communists were not, and it was they who took control of the strike and set it on a course that would lead to a break with the AFL.

The Northwest Joint Strike Committee

Abe Muir hoped to negotiate a uniform working agreement with the major Longview lumber operators that could be applied to the entire lumber industry in the Northwest,[63] but on May 10, about nine hundred Long-Bell and Weyerhaeuser Lumber Company employees in Longview walked out of a union meeting before a vote could be taken to ratify an agreement Muir had negotiated with company representatives. On May 13, the same loggers and sawmill workers voted nine to one to reject the May 10 proposal of fifty cents an hour, a forty-hour week, time-and-one-half for overtime and union recognition with no closed shop. Militant unionists soon referred to this proposal as the "sell-out" or "Muir Compromise."[64]

The Longview negotiations served only to confirm dissatisfaction with Muir. Since May 3, the *Voice of Action* had been demanding his resignation and the formation of a Joint Strike Committee to establish rank-and-file control of the strike. At Aberdeen on June 5, 240 insurgent anti-Muir delegates from the Sawmill and Timber Workers Union and the dormant National Lumber Workers Union met to wrest control of the strike from the Carpenters' executive. They rejected Muir's Longview settlement and formed their own Northwest Joint Strike Committee to open new negotiations with the lumber operators.[65]

Forty locals were affiliated with the Joint Strike Committee by July 1. On June 9, the Aberdeen boom workers had voted to support the Committee; sawmill workers there voted 970 to 57 on June 23 to continue on strike; three days later lumber workers in Portland unanimously voted to stay out for a guarantee of collective bargaining. By 1,500 to 3, Everett, Washington lumber workers rejected a back to work offer on July 11; only 51 out of 1,157 workers wanted to accept Muir's proposal at Tacoma on July 28; two days later, the same workers refused to even reconsider the proposal. On August 1, Raymond, Washington lumber workers voted nearly two to one to reject Muir's plans. The votes clearly belied the media's picture of the strike as the work of a handful of radicals in the Northwest Joint Strike Committee.[66] Communist lumber workers did, however, participate in the Joint Strike Committee.

On March 16, 1935, the Trade Union Unity League was dissolved and the Communist Party once again adopted the policy of "boring from within."[67] Communist labor organizers were no longer committed to establishing "dual unions" such as the National Lumber Workers Union, which tended to be sectarian and isolated from the mass of lumber workers.[68] They were now free to organize within existing AFL unions. As early as February 23, 1935, the *Voice of Action* carried a statement from the Northwest District Communist Party that urged party members to join the AFL and

promote concrete demands such as fights against wage cuts, rising pay for rising prices, a reduction in hours without reduction in pay, elimination of the speed-up, real social insurance and unemployment relief.[69]

On April 15, 1935, the Communist-led National Lumber Workers Union officially disbanded. During a conference at Tacoma the union's leaders proposed that the entire membership join the AFL's Sawmill and Timber Workers Union ". . . and make the AFL more militant despite the AFL administration." Shortly thereafter, James Murphy, former president of the National Lumber Workers Union and an active member of the Communist Party, joined the Northwest Joint Strike Committee and began exercising considerable influence in its deliberations.[70]

Communists brought years of organizing experience into the movement. Karly Larsen, a Danish immigrant working at Lyman Timber Company, had joined the IWW at thirteen and had been organizing among the local unemployed since 1930. In one action Larsen and others had led a hunger march of seven thousand to the state capital in Olympia; in another, they led a raid on an Anacortes grocery store and redistributed food to the unemployed. Shortly after the strike, Larsen was elected to the District Executive Board of the AFL's Sawmill and Timber Workers Union.[71]

The tactics of the communists and left-wingers in the Strike Committee were similar to those employed during the 1934 Longshore Strike. They placed Party members on the picket lines whenever possible and distributed leaflets to persuade union members to support the Committee and the original demands. Almost every night and on weekends Committee members met to discuss strategy and establish the evening's "roving committee." These special groups within the larger Joint Strike Committee were charged with the responsibility of bolstering picket line discipline and preventing a back-to-work movement. Under cover of night, carloads of men moved into mill towns from Portland to Aberdeen and north to Everett. With few mimeograph machines and little access to radio, it was imperative that the roving strikers meet with other rank and filers to combat Muir's steady "compromise" propaganda.[72]

The decision of the National Lumber Workers Union to disband and "bore from within" the AFL had marked results as the Northwest Joint Strike Committee and its activist "roving committees" pushed lumber workers to the left and toward industrial unionism.

Whenever prior contact with Unemployed Councils or with "Muir's compromise" failed to move a logger to support the strike, the roving committees advocated other methods. At a logging camp in Oregon's North Coast Range a rigging slinger who had been wavering in his decision to join the Strike Committee was spied up a spar tree that had a hornets' nest propitiously nestled at its base. Chunks of bark heaved from a safe distance by fellow loggers soon shattered the nest, and the confused hornets quickly attacked the rigger. After repeated confrontations with the hornets, the reluctant logger agreed to sign up: "I'll join your union if you'll only let me down."[73]

But Muir was a different matter. When he realized that confrontation with the Strike Committee threatened the future of the Sawmill and Timber Workers Union, he countered the lumber workers' solidarity by expelling those union leaders who opposed his "compromise" settlement. Those lumber workers who still rejected his leadership, he punished by revoking their local union charters and summarily placing them in separate locals under his jurisdiction. At the same time, many small locals, as well as larger units in Bellingham and Port Ludlow, Washington, and Coos Bay, Oregon, remained loyal to him. The powerful and conservative locals in Portland were also his consistent supporters.[74]

The lumber operators also joined with Muir and the Carpenters' officials to denounce the Communists and left-wingers who encouraged workers to hold out for the original strike demands. At the early negotiations with the Long-Bell and Weyerhaeuser lumber companies, Abe Muir cautioned the employers that a "flying squadron of Reds" was coming up from Portland to picket the mills. The AFL's *Oregon Labor Press* had previously warned mill operators to ". . . deal with the regularly organized AFL union in the industry" or face the possibility of "dealing with Communists in the future."[75]

Muir's plan was simple: he expected the lumber workers to accept his strike settlement when given the alternative of dealing with a Communist-dominated Strike Committee. The lumber operators hoped to prevent a successful conclusion to the strike by identifying all the activists in the Committee with the Communist Party. To the dismay of both Muir and the lumber companies, the strategy failed to work; the majority of lumber workers refused to desert the Strike Committee,[76] and they disregarded the accusation that it was completely Communist-led. They contended that the Committee had been formed to "...free the membership [of the union] from [Abe Muir's] dictatorship." But while the Committee wanted Muir removed immediately, it had no desire at that time to withdraw the lumber workers from the AFL and form a new, Communist-led union.[77]

The striking lumber workers got help from the Unemployed Councils and from President Roosevelt's Federal Emergency Relief Administration which granted relief benefits to lumber workers on strike. Since the International Longshoremen's Association had a debt to repay from the 1934 longshore strike, the ILA's members refused to load lumber produced by members of the Muir faction. One well-known Communist longshoreman, Harry Pilcher, took leave from the ILA to organize for the Northwest Joint Strike Committee. He advised loggers and mill workers to adopt the militancy and unified stance that had enabled the longshoremen to win. Many of the most militant loggers' locals along the Northwest waterfront were in close communication with the longshoremen.[78]

But employers could also marshal support. Assessments were levied to obtain the funds necessary to defeat the strike. In Seattle, the "Committee of Five Hundred," which the Strike Committee believed was really a front for the Chamber of Commerce, placed full-page ads in the newspapers. According to this "Committee," the disagreement was all the work of a few burly, radical loggers who recklessly continued the strike even though it meant ruination for the industry. Daily news bulletins went out to Northwest lumber operators from the Portland office of *Crow's Pacific Coast Lumber Digest*, the industry's West Coast publication. The editor's predictions of a quick failure for the strike helped solidify the resistance of the employers. In May, 1935, he announced an "...inevitable collapse of the strike...," on May 21: "...each day brings added reports of an increasing number of employees expressing a desire to be back to work on the *old basis*" (emphasis added); May 25: "...strike leaders are not admitting that they are whipped." And on June 15: "...the strike is well on its way out."[79]

State and Employer Violence

The demand for containers to ship the Northwest's fruit and vegetable harvest increased the pressure to end the strike by late June. Lumber to fill the orders was either not being produced or being boycotted on the docks by longshoremen who refused to handle "hot cargo." The first use of violence by the employers and the police came in Humboldt County, California.[80]

Northern California mill owners who blamed the strike on outside agitators had responded by hiring a private militia to help local police; they had laid in a supply of tear gas and enthusiastically recruited strikebreakers. On June 21, 1935, Eureka, California police and gunmen hired by the mill owners opened fire on unarmed pickets assembled at the entrance to the Holmes-Eureka mill. Three men, all Finns, were killed. Scores more, suffering from gunshot wounds and exposure to tear gas, were taken to the hospital. Others, fearing arrest, were carried home to be treated by family and friends. For days afterward, lumber workers who supported the strike were arrested and jailed on assault charges for strike activities.

On June 26 trouble began nearer to the strike's centre of strength when Washington Governor Clarence Martin promised either State Police or National Guard protection to all employers who wished to reopen their mills. His radio address was laden with stereotypes gleaned from the employers' propaganda and from the headlines of Seattle's large dailies. Strikebreakers were depicted as "workers," union men as "agitators." According to the Governor, the majority of lumber workers wished to return to work, but they were intimidated by a few radicals. Strikers and their sympathizers would no longer be allowed to terrorize workers who otherwise would have returned to work long ago.

Soon, the inevitable confrontations and fights between striking pickets and "scabs" began. The Strike Committee contended that police and guardsmen precipitated the violence when they sheltered those who were trying to break the picket lines. Pickets beat "scabs"; strikebreakers beat pickets. It depended upon who had the upper hand at the entrance to the mills and logging camps.

At Longview, Washington state police fired tear gas to scatter the pickets. When the strikers obtained an injunction to halt the intervention, the state police paid little attention to the court. In Aberdeen and Hoquiam, Washington on July 7, National Guard troops and State Police reinforced local police who had already clashed several times with the two thousand pickets who were blocking the employers' attempt to open the mills. The next day, six to ten thousand demonstrators marched through the "Twin Cities" to protest Governor Martin's use of troops and police to provide protection for the strikebreakers. Rumors circulated that a general strike would be called if the soldiers were not withdrawn. But Martin was steadfast; the mills would be reopened for those who wished to work. He continued to profess that the strike was mostly the work of Communists.

In Tacoma, the strikers clashed repeatedly with guardsmen who protected the movement of scabs to and from the mills. They had received orders to prohibit picketing, and to prevent pedestrians from assembling in the industrial part of the city. On June 12, seven thousand union men and strike sympathizers marched peacefully in a large circle at the downtown intersection of Eleventh and "A" Streets, their leaders carrying American flags. Although the marchers used the crosswalks, traffic was blocked. The site of the march had been chosen because it was adjacent to the Eleventh Street bridge which connected the city centre with the industrial section of Tacoma, the same bridge that the strikebreakers had to pass over every day on the way home from work. Ready for trouble, police stood by and guardsmen with fixed bayonets established a line across the bridge.

Taunts and jeers greeted the strikebreakers as they came down the bridge. Women and children joined in the booing and helped pelt the guardsmen with rocks and sticks. The troops lobbed tear gas bombs into the crowds, but just as quickly, the wind blew the smoke back into their own faces. Other tear gas cannisters were picked up by the marchers and hurled back amid the guardsmen, only some of whom had appropriate gas masks. The police then called in heavier equipment. A military vehicle which expelled tear gas through its exhaust pipe was sent into the crowd but its driver was pulled from the cab and attacked before the overturned truck caught fire and burned in the middle of Eleventh Street. After that, the troops abandoned the bridge and the last significant battle between lumber workers and guardsmen was over. The demonstrators soon dispersed but arrests followed.

While State Police and guardsmen confronted strikers on the outside, Abe Muir was busy on the inside, sowing intra-union dissension. In Aberdeen and Everett, previously tested strikebreaking tactics were again applied. Muir revoked charters of locals that continued to support the Northwest Joint Strike Committee and granted new charters under leadership from outside the Committee. Undercut by both Muir and Martin, the Committee appeared to have lost the strike.

In Portland, meanwhile, the Northwest Joint Strike Committee and Communist activists had made little progress. Party members who were actively involved with the Committee occasionally came down to Portland to encourage the sawmill workers to support the strike, but President William Wedal and Secretary Frank Johnson of the Portland Sawmill and Timber Workers local both supported Abe Muir. Johnson vowed that local sawmill workers were standing ". . . one hundred percent behind the American Federation of Labor boys."[81] The mill owners refused to recognize the Strike Committee and preferred instead to negotiate with Muir's more conservative representatives. Since they numbered less than 300 of the 2,500 local members, the Strike Committee in Portland could therefore not press for a settlement based on their original demands.

The inability of either the strikers or the employers to win an outright victory opened the way for federal mediation. Franklin Roosevelt's Secretary of Labor, Frances Perkins, appointed a three-member Federal Labor Mediation Board late in June. The compromises inherent in the process were acceptable to the lumber workers, but not to Governor Martin and the employers.[82] Why should lumber companies submit to arbitration, they asked, when the strike was bound to fail? They were, moreover, suspicious of a labor board appointed by a Democratic president's Department of Labor. But their rejection of the Board was couched in terms calculated to place the onus on the Northwest Joint Strike Committee for the continuation of the dispute: "It is no longer a question between employers and employees, but between the constituted authorities and a lawless element."[83]

By late July, 1935, the strike had, for all practical purposes, collapsed even though employers continued to exaggerate the numbers of workers willing to return to work. The lumber operators implemented Muir's agreement: lumber workers should be willing to work for fifty cents an hour minimum for a standard forty hours per week. Mills in Seattle, Bellingham, Aberdeen, Portland, Longview and Olympia were soon working under these terms. Instead of a "closed shop" and strict union recognition, employers agreed to acknowledge committees of union members authorized to bargain over safety and working conditions in the camps and mills.[84] Disagreements and scattered clashes continued, especially where lumber operators failed to rehire former radical unionists, but militant locals were not above calling a "quickie" strike to insure that a strike activist was re-employed.

On August 5, Tacoma lumber workers finally voted 1,391 to 97 to accept the settlement, but Grays Harbor militants held firm and State Police continued to confront the pickets there. Then on August 13, Governor Martin appointed Washington National Guard Brigadier General Carlos Pennington to act as "mediator."[85] In less than one week, the strike in the harbor towns of Aberdeen and Hoquiam was over; by then, most lumber workers throughout the Northwest were back in the mills or on their way to the logging camps. Operations were once again at "full-bell."

Six months earlier, the lumber workers had drawn up their initial demands: a thirty-hour week; seventy-five cents per hour minimum; union recognition; paid vacations; seniority rights; and time-and-one-half for overtime. After a tough three-month strike, they were forced to accept Muir's compromise: ten hours more work per week and twenty-five cents less an hour than the original demands, no union recognition, and little else. From a pre-strike membership of roughly 15,000 to a membership of 35,000 at the strike's conclusion, the union continued to expand until, by the spring of 1936, 70,000 lumber workers belonged, making it the largest body of workers on the West Coast. But it had grown not because of Abe Muir's leadership, but in spite of it.

Insurgents of the Woods and Mills

Dissension plagued the union. Abe Muir, a centre of controversy during the strike, continued to alienate lumber workers after the strike ended. His craft-union bias separated him from the militant loggers and sawmill workers laboring under the "compromise agreement" he had imposed on them. Muir and other Carpenters officials, however, were emboldened by the failure of the strike, and proceeded resolutely with their efforts to control the lumber workers and inhibit the growth of industrial unionism in the Northwest woods.

Their first move was against the Northwest Council of Sawmill and Timber Workers, AFL, whose annual convention was scheduled for Centralia, Washington, on October 12, 1935. When Muir discovered that insurgent lumber workers from thirteen locals had held a pre-convention meeting in Olympia in the first week in October to "...plan an attack on Muir's leadership," he postponed the convention and rescheduled it to open in Portland a week later. The insurgents alleged the convention was shifted to distract and confuse the opposition to Abe Muir and the United Brotherhood. When the convention finally opened, Muir promptly dissolved the Council and reorganized Sawmill and Timber Workers into small district councils located in the major lumber centres of the Northwest. According to Muir, the new district councils were created because the Northwest Council had served its function as a body of federal unions and was no longer necessary for organizational activity. In reality, the new district councils were formed to enable Muir to exert more control over the insurgent lumber workers who had dominated the Northwest District Council of Sawmill and Timber Workers and had opposed him during the recent lumber strike.[86] From October, 1935, following the dissolution of the Northwest Council, to September, 1936, opposition to the union leadership of the Abe Muir faction continued to grow.

In British Columbia there were new rumblings afoot. Since the 1934 lumber strike there, a fragile truce had settled over labor relations north of the border, and although the great Northwest Strike of 1935 had elicited much discussion, it had never been extended into western Canada. Then in April, 1936 the Communist-led Lumber Workers Industrial Union balloted 1,048 to 23 to merge with the UBCJ to become the Lumber and Sawmill Workers Union of British Columbia, headquartered in Vancouver.[87] After that, in both the Northwest and British Columbia, Communist lumber workers could organize within the mainstream of the trade union movement. Their goal was a "United Front" of lumber workers and carpenters who would oppose the rising tide of war and fascism in Europe and stand for industrial unionism within the AFL.

The first fruit of this new alliance was a short-lived strike in British Columbia in May and June, 1936. The UBCJ District Council pulled out 2,000 of the province's 20,000 loggers and sawmill workers. It was not enough to mount a successful general strike in the industry, but it did garner a fifty-cent hourly wage increase and expanded the membership of the Sawmill and Timber Workers Union in British Columbia. During this strike, a loggers' trek from Cowichan Lake Camp to Campbell River was planned to protest the summary firing of two camp stewards and to galvanize strike support. The arbitrary cancellation of the trek forced the union president to resign and Harold Pritchett was once again cast centre-stage as the new president of the UBCJ's District Council in British Columbia.[88]

At Centralia, Washington, representatives from the British Columbia, Northern Washington, Grays Harbor-Willapa Bay, Columbia River, Coos Bay and Klamath Falls district councils gathered in late June, 1936 to discuss the differences between Muir and the lumber workers. They adopted resolutions which affirmed that the Sawmill and Timber Workers Union would remain an "industrial union" and continue to resist the encroachment of craft unionism on the lumber industry. A committee of fifteen, elected from the union's district councils, was invested to "...make a thorough study of the [industrial union] situation and to work in conjection [sic] with the national [Committee for Industrial Organization] CIO." Further, the UBCJ must reverse its "tyrannical" decision to grant only fraternal status instead of full voting privileges to Sawmill and Timber Worker delegates at the Carpenters' conventions. If not, lumber workers would "...consider what further action is necessary to secure for us our just rights and privileges as members of the United Brotherhood of Carpenters and Joiners." This would be done in September at the special lumber workers' convention which the Centralia representative had called in order to resolve the strife with Muir and the United Brotherhood.[89]

The Federation of Woodworkers

Six hundred and twenty-five delegates from ten district councils answered the summons to meet in Portland, Oregon on September 18, 1936. The convention was packed; lumber and sawmill workers, plywood and veneer workers, pile drivers, coopers, shingle weavers, furniture workers, and boommen and rafters, all were represented. The rebellious delegates promptly reorganized the lumber workers and formed the Federation of Woodworkers. Neither chartered nor recognized by the United Brotherhood of Carpenters and Joiners, the Federation was an unaffiliated organization superimposed over the new district councils Abe Muir had created.[90]

Practically overnight, the Federation of Woodworkers, with 72,000 members, became the largest union organization on the West Coast. Its goal — a united, industrial union of lumber workers — resembled the old plan of the old Northwest Joint Strike Committee. In a little more than a year, the spirit of the '35 lumber strike had been rekindled. To make matters worse for Abe Muir and the Carpenters, Harold Pritchett now staked a claim to the leadership of the Northwest lumber workers as well; he chaired the convention and was later elected to the presidency of the new Federation of Woodworkers.

Despite the tone of the convention, the Federation of Woodworkers wanted to work within the UBCJ, and still hoped to change the attitude of the UBCJ towards industrial unionism for the lumber workers. The defeat of a resolution to withold the Federation's per capita tax from the Carpenters Brotherhood indicated that a majority of the delegates did not want to break with the Carpenters.[91] However, both Abe Muir and William Hutcheson refused to countenance any change in the organizational policies of the Carpenters which might pacify the Federation of Woodworkers.

Three months later, sixteen delegates from the Federation of Woodworkers stepped from their chartered Pullman to attend the United Brotherhood of Carpenters and Joiners Convention in Lakeland, Florida. Their demand for full voting privileges on all matters connected with the lumber industry was snubbed; the lumber workers were seated as fraternal delegates and denied the right to vote.[92] Frank Duffy, general secretary of the UBCJ, leaned from the rostrum and declared the lumber workers undesirable as union members. The Brotherhood would not organize the lumber workers into industrial unions, and if they dared to form a "dual union," he vowed that the Carpenters would give them ". . . the sweetest fight of their lives."[93]

After the convention, the disgruntled delegates headed north to Washington, D.C. to confer with John L. Lewis, president of the CIO, and John Brophy, the CIO's Director of Organizing. At the time, Brophy and Lewis were leading the CIO organizing campaigns in the steel and automobile industries, and it was perhaps because of this, or because they lacked adequate funds, that they failed to give the delegates from the Federation of Woodworkers much encouragement. Since the Federation of Woodworkers was still associated with the United Brotherhood, it did not require the immediate attention which the "unorganized" condition of the auto and steel industries demanded.[94]

Delegates to the second convention of the Federation of Woodworkers met in Longview, Washington from February 10 to February 22, 1937. There they considered the future policies of the organization in light of the indifferent attitudes of the AFL and CIO executives.

On August 4, 1936, the AFL Executive Council had suspended eight unions affiliated with the Committee for Industrial Organization, and early the following year, had expelled them. Consequently, the first resolution at the February convention of the Federation of Woodworkers, unanimously adopted, urged the AFL Executive Council to "reinstate those union members recently suspended for affiliating with the new Committee for Industrial Organization." The Convention then endorsed the principle of industrial unionism and requested the AFL to lend ". . . every assistance both morally and financially to the Committee for Industrial Organization." Another resolution reflected the ill will generated by the 1935 lumber strike; it called for the removal of Abe Muir as international representative of the Carpenters Brotherhood.[95] Finally, the convention adopted the recommendation of the executive board that the lumber workers remain in the AFL in the hope that the continued alliance would at least aid the Federation's new organizing drive.[96]

Other resolutions passed by the convention placed the Federation of Woodworkers politically to the left of both the United Brotherhood and the AFL. The convention sent a "communication of support and unity to the Spanish trade unions in Madrid," and urged the locals to contribute to the Trade Union Fund of the Spanish Labor Red Cross. Another resolution called for the ". . . universal conscription of property and the elimination of profit through all industry in time of war." Franklin D. Roosevelt was unanimously endorsed in his 1936 bid for re-election, while William Hutcheson, president of the United Brotherhood of Carpenters and Joiners, was rebuked for his support of Alf Landon.[97]

The Communist leaders of the Federation offered lumber workers an alternative to the do-nothing kind of union leadership represented by the Abe Muirs of the AFL. "Delivering the goods" in the depression years meant organizing. Workers wanted unions and they wanted employer recognition of those unions. By working in the Unemployed Councils and the Trade Union Unity League Affiliates, Communists and those who would soon join the Communist Party established their credentials as capable organizers.

The success of the Communists as organizers cannot be attributed to personal qualities of individuals like Karly Larsen or Harold Pritchett.

While it is true that they were personable people and hard workers for the union, they were probably no more so than men who had tried and would try to organize with less success. The Communists were successful because they had an understanding of the class relations that lay beneath the crisis of the Great Depression and were able to translate that understanding into successful strategy and tactics. They were the individuals who responded to the needs of the lumber industry communities left destitute by the Depression and they were able to use their ties to the community and to community resources to enhance the organizational capacity of the fledgling union movement. Finally, their ties through the Communist Party enabled them to bridge the separations between communities, between industries such as wood products and longshoring, between political, economic and cultural organizations and across national boundaries. In a struggle such as the one which was unfolding in the wood products industry working class unity in the widest possible dimensions would be necessary for success, and by 1937 the difference between the practice of Communist unionists and that of the AFL's business unionists had become apparent.

3
Toward One Union in Wood

Historical circumstances had insured that industrial unionism for wood-workers would not be easily achieved. By 1937 the movement's leaders were already fighting on four fronts. The employers were determined to do everything possible to prevent industry-wide worker unity; the AFL had a major stake in keeping out a competitor; the state, on both local and federal levels, was closely aligned with employer interests; finally, conservatism within the movement's own ranks constituted a "fifth column". This combination would eventually prove decisive in defeating the union's left-wing leadership.

The 1937 Midwest Lumber Strikes

Prior to 1937 all attempts to organize woodworkers in the Midwest had ended in failure. In the 1920s the Socialist Finns who joined the Industrial Workers of the World had carried the union banner but never succeeded in forming a permanent body to represent the timber workers. In the early 1930s the Communists' National Lumber Workers Union had also tried and met with defeat. Then in the fall of 1935 Minnesota had its first lumber strike in twelve years, and although it failed, the Carpenters and Joiners consented to grant a charter to the Minnesota lumber workers in April of 1936.[1]

In January of the following year, the Northern Minnesota lumber workers struck again. Farmer-laborite Governor Elmer Benson declared it a just strike and, in an unprecedented move, ordered both the National Guard and the Highway Patrol to aid the strikers. In Duluth the Armory was heated and opened for homeless lumberjacks, and the Guard

47

distributed blankets to strikers and helped organize soup kitchens at several strike locales. By January 18 Minnesota Local 2776 of the UBCJ's Lumber and Sawmill Workers had reached a settlement, and a six-month contract with improved wages and working conditions was signed with 192 logging companies. Union membership climbed to over 5,000 and the local's newspaper, the *Timber Worker*, began weekly distribution. More significantly for the Lake States, left-wing Minnesota lumber workers shunned isolationism in their victory, setting their sights instead on organizing the entire logging industry from Minnesota right on across Wisconsin to Michigan.[2]

The big push came in Michigan's Upper Peninsula. The UP, a wedge of Michigan sitting atop Wisconsin and cut off from the rest of thê state by Lake Michigan, was at that time dominated by five giants of the Lake States' lumber industry. The Cleveland Cliffs Iron Company owned and logged 371,673 acres of Michigan timber. They also had interests in pulp and paper, mining timbers, railroad ties, household utensils, iron mines and the Great Lakes ore trade. They were the leading corporate interest on the Michigan Upper Peninsula, yet their wages in 1932 for their three hundred loggers averaged $12 to $16 a month. The Newberry Lumber and Chemical Company in the eastern UP employed 325 workers, not counting those in its lumber camps. The William Bonifas Company at Marenisco, a subsidiary of the Kimberly-Clark Paper Company to whom it supplied pulp wood, controlled the western edge of the Peninsula.[3] Operations belonging to the Ford Motor Company employed nearly one thousand workers at the mammoth Iron Mountain complex in the south central UP. The hardwood, which became floorboards and panels and numerous other components of the 1930s Fords and "Woodie" station wagons, was transported the short distance to the assembly lines at the Ford plant in Kingsford.[4] The fifth giant, the Connor Lumber and Land Company, which was destined to become one of the world's largest manufacturers of furniture and hardwood flooring, was headquartered in Laona, Wisconsin, thirty-five miles southwest of the Ford plants. It was a relatively short trip on the C & NW railroad to the company's main UP logging area in Gogebic County. There, in the company town of Connorville, two hundred workers labored at the mill manufacturing the timber cut by another two hundred lumberjacks in the Connor woods operations.[5]

In March 1937, the mostly Finnish Communist Party in Ironwood, Michigan aided by Communist leaders of the Minnesota strike, joined with local lumber and sawmill workers and applied successfully for an AFL charter.[6] The new local received help from Minnesota in the form of Joe

Liss, a well-known organizer who had been wholly responsible for publishing the first six issues of the *Timber Worker*. Then on May 11, half the crew at the William Bonifas Camp II walked out. At a mass meeting staged at the Marenisco Town Hall, leaflets were handed out listing Local 2530's demands: fifty-five cents an hour, a forty-hour week, single beds instead of two-man bunks, showers, and recognition of both the union and camp communities.[7] Only half in jest, Matt Savola, a young Communist Party section organizer sent to Marenisco to help the strikers, coined the rallying cry: "Lumberjacks, unite! All you've got to lose is your bedbugs!"[8]

From the beginning the strike was led by radicals. Some had worked previously with the Worker's Alliance, organizing in Roosevelt's federal programs; others had gained experience in the farmer's co-ops that stretched across the Lake States. But a composite biographical sketch of these union organizers would probably show a second generation Scandinavian — often a Finn — whose family had moved to the Lake States to work in the mining or lumber industries. For example, George Rahkonen, the Finnish secretary-treasurer of Local 2530, recruited Luke Raik into the Communist Party, and Raik, a Marenisco lumberjack of Austrian descent, became Local 2530's first president. Matt Savola, a Communist and a Finn, was chairman of the Strike Relief Committee. Radical attorney Henry Paull, Irene Paull (who wrote under the pen name "Calamity Jane" in the *Timber Worker*), Joe Liss and dozens of others who counted themselves as part of the left-wing movement joined the strike effort.[9]

The Michigan lumber workers' strike spread quickly across the peninsula. By June 4, eight strike offices had been opened and 4,000 lumber workers were out. Four of Ford Motor Company's six camps were closed.

At 6:00 a.m. on the 4th, workers from the Newberry Chemical and Lumber Company assembled and began marching two abreast toward the plant. The company fire whistle blew, bringing townspeople, foremen, night shift employees, the plant manager (who was also the Newberry village president), and the local sheriff — 300 people in all — to the scene. When the strikers tried to withdraw, the mob attacked with clubs, gate handles, iron bars, rubber hoses, hose couplings and iron bolts. The strikers were chased for three miles and one worker was killed. Later, the mob ransacked the Finnish Workers Hall, which had been an organizing centre for the strike and beat those unionists who had sought refuge there. That same afternoon, Joe Liss led a march to the Munising Courthouse to demand relief for the strikers. He and another organizer, David "Double Breasted Joe" LeClaire, were arrested.[10]

Henry Paull, the union's lawyer, arrived the next day. He already had two strikes against him when he appeared at the office of the justice of the peace. First, he was a Jew and the son of Russian immigrants from Kiev. Second, he had defended two Communists who had been arrested and jailed three years earlier for having raised the Red flag in Eben Junction, a small town not far from Munising. As Paull and his wife Irene approached the courthouse with union president Luke Raik and *Timber Worker* editor Sam Davis, a crowd of 200 gathered, led by Sheriff Pelletier and the prosecuting attorney. Lead pipes and clubs appeared. Slowly the mob forced the four to retreat down the street.

"We'll get you, you Jew!"

"Communist!" they yelled.[11]

A shopkeeper who was patronized by lumberjacks sheltered them briefly in his doorway but the mob kept coming. Finally, when the four reached Paull's car, they were escorted out of town to the county line. Not satisfied, police and loyal townspeople rounded up the remaining union members and hurried them out of town too.[12]

Within two days, vigilantism won out in Newberry and Munising and a "back to work" movement was underway. But in the western tip of the Peninsula, around Marenisco, Ironwood and Bessemer, the union refused to be so easily dislodged.

In that region, the union men controlled the roads, and they set out to stop the truckers or "jobbers" from hauling the timber to the mills that the loggers had cut that spring. If the sawmills could operate while the loggers were on relief, the economic burdens would be unequal and the companies could more easily wait out the strike. The workers' tactics were simple: when trucks geared down for a steep grade, a striker from the crowd of several hundred pickets that lined both sides of the road would jump on the running board. If a brief conversation with the driver failed to halt the truck, stones through the windshield usually did. Loads were dumped, one load becoming free firewood for Marenisco's citizens when a striker unloaded the logs in someone's yard. Even with state police and special deputies on patrol, few log trucks were able to deliver their loads to the mills. Unfortunately, these tactics gave the jobbers a perfect pretext to call off negotiations with the union.[13]

On the night of June 30, vigilantes gathered at the Gogebic County Fairgrounds, then swept down on the North Ironwood strike headquarters at the North Star Hall. The Swede-Finn Hall, strike headquarters in neighboring Bessemer, was next. That night, Henry Paull, Luke Raik and Jim Rodgers, vice-president of the Minnesota Timber Workers, were meeting in Ironwood with Nathanial Clark from the National Labor Relations Board.

After their meeting, the three union men walked to a cafe across the street. Shortly afterwards, the vigilantes arrived from Bessemer, pillaged the union headquarters, then burst into the cafe. Paull and Raik were dragged outside and savagely beaten with axe handles. A car narrowly missed Paull as he broke free and dashed down Ironwood's main street in full view of the police, who stood looking on. The vigilantes continued to hound the lawyer until they trapped him, grabbed him by the legs and dragged him by the feet to a waiting car.

Paull and Raik's captors discussed killing them, but decided better; instead, they drove them away from Ironwood in separate cars and dumped them across the border from Michigan's Upper Peninsula in Saxon, Wisconsin. After he was found by his wife and a group of strikers, Paull spent the next three days in the hospital.

But the vigilantes' appetite was still not satiated. Later that same night and all the next day, they worked hard to fulfill the threat a jobber had hurled at Paull during his car ride to Saxon: "We are going to finish [those] bastards [union men] tonight so that they won't be able to interfere with us anymore."[14] Palace Hall, the Ironwood strike headquarters, was forced open and its occupants were subjected to another gauntlet before being driven out of town. The next day police and vigilantes continued to round up strikers; organizers' cars were stopped and confiscated, and the beatings continued. In all, sixty-four striking lumber workers or their supporters were arrested.[15]

Following the violence at Newberry, Munising and Ironwood, Michigan's governor Frank Murphy was swamped with protests, including a strongly worded message from Governor Benson of Minnesota, since some of that state's citizens were in need of protection in Michigan.[16] Murphy appointed a committee of five to conduct an investigation. On July 14, the committee held an election at the William Bonifas Camp II to evaluate support for the union, and concluded that a majority of the workers who were employed when the strike was called had willingly supported the strike action. However, agreements were only reached with a few small operations.[17]

Throughout the strike, the AFL and the Carpenters' Brotherhood had been hesitant participants because they were reluctant to follow the lead of the leftists who headed the strike movement. The AFL's Joseph Ashmore, a long-time member of the craft-dominated Painters' Union, was Governor Murphy's Secretary of Labor and Industry; time and again he had attempted to sell out the strikers. When Henry Paull was chased out of Munising, Ashmore had been there in conference with his own hand-picked representatives from the union. Paull requested that Ashmore provide police protection for himself and Luke Raik so that they could return to Munising to participate in the negotiations. Instead, Ashmore red-baited the strike's leaders: "We will assist the working men to organize in a proper manner", he announced. "Their present organization [has] adopted methods that even Soviet Russia would not tolerate."[18]

Meanwhile before Michigan's lumber workers had ended their strike, the national movement of their brothers and sisters to join the CIO had reached a conclusion. On July 15 the Federation of Woodworkers met in Tacoma, Washington and five days later emerged as the International Woodworkers of America — CIO. The AFL's Michigan Local 2530 reorganized itself as Local 15 of the IWA and within months joined with other locals in Minnesota and Michigan to form Midwest District Council 12 of the IWA. As a consequence, when the strike was settled in August, it was the members of the IWA, not the UBCJ's Lumber and Sawmill Workers, who went back to work.

Matt Savola became the first president of the new Local 15 and the first secretary-treasurer of the District Council. George Rahkonen was a member of the executive board along with Ernest Tomberg from Minnesota. It was the radicalism and cohesiveness of its Finnish members that kept the Midwest District from splitting into pro- and anti-communist blocs.

The Conflict With the AFL: 1937

While Michigan lumber workers were striking, the Federation of Woodworkers had met on June 7 and 8 to consider the possibility of affiliating with the CIO. William Hutcheson, president of the United Brotherhood of Carpenters and Joiners, ignored an invitation from Pritchett to attend the conference and instead sent Abe Muir. However, John Brophy, CIO Director of Organization, and Harry Bridges, President of the International Longshoremen's Association and West Coast organizer for the CIO, did accept the invitation to attend. They addressed the delegates, inviting them to affiliate with the CIO, and Brophy offered the lumber workers financial support and a national CIO charter.[19]

On the morning of June 8, a resolution to submit the question of CIO affiliation to a referendum vote passed the conference with a fifty-six percent majority. The results of the referendum would be considered at the Third Convention of the Federation in Tacoma, Washington, on July 17. The Columbia River District Council, which controlled all the workers on the Oregon side of the river from the Dalles to Astoria and down the coast to Coos Bay, considered this referendum resolution to be premature and felt the Council was being "railroaded" into joining the CIO, but the Federation's leadership found this hesitancy unwarranted as it had been nearly a year since the question was originally raised. During that year the UBCJ had remained insensitive to the lumber workers' appeals for industrial unionism and had again proven itself incapable of leading strike action. Although they voted five to two against the referendum resolution and opposed Pritchett's leadership, the Columbia River District Council remained in the Federation.[20]

To head off the burgeoning CIO movement, Hutcheson belatedly ordered all the former executive board members of the Northwest Council of Sawmill and Timber Workers to meet with him in Longview, Washington on June 19. Hutcheson's men appeared for the conclave, but so did three or four hundred uninvited Federation members, and the overflow meeting had to be moved from the hotel dining room to the auditorium of the public library. When the meeting had reassembled, the Carpenters' president called for silence, then threatened the lumber workers with a UBCJ boycott of all CIO-produced lumber if they dared to enter any new CIO union. When the lumber workers retaliated by demanding industrial union representation, Hutcheson accused them of being a "pack of Communists."

"I am not here to grant you anything, we're not here to give away anything. We're here just to hear what you have to say, and that's it."

His pronouncements so antagonized the assembled lumber workers that their shouting forced him and his supporters to abandon the stage. Harold Pritchett picked up the gavel and called the meeting to order. After a brief, enthusiastic discussion, the remaining lumber workers decided to proceed with their drive to "Go CIO."[21]

Harold Pritchett, John Brophy and O.M. "Mickey" Orton, a member of the Federation of Woodworkers' executive board, traveled to Washington, D.C. during the first week of July to confer with John L. Lewis. Although the referendum votes would not be tallied for several weeks, the Federation's leaders expected a CIO victory and wanted to complete the arrangements for a CIO Charter.[22]

Given the union members' opposition to the UBCJ, the Federation of Woodworkers' Third Convention which convened in Tacoma on July 17 was a surprisingly dissentious affair. In fact, the length of the debates extended it from one to three days. The affiliation referendum ballots had been deposited in sealed boxes and then counted by a committee which included one member from each district council executive board. When it was announced that the vote favored CIO affiliation by 16,754 to 5,306, some delegates charged the executive board of the Federation with having stuffed the ballot box. However, there was no evidence of "stuffing." For example, employees at the M&M Woodworking Company in Portland had held a secret ballot and sent the results to the Tacoma convention in sealed ballot boxes. When their votes were counted, eighty-nine percent of the local had voted for CIO affiliation.[23] Other delegates charged that since only one-fourth of the Federation's membership had voted, the ballots favouring the CIO only represented a minority, but it had been the anti-CIO leaders in some districts who had refused to allow their members to vote on the issue.

On the fourth day of the convention, delegates supported Harold Pritchett's call for immediate affiliation with the CIO by a vote of 385 to 71. Then a debate ensued over what to call the new union. The International Wood Workers of America was a popular choice, but the initials I.W.W. of A. smacked too much of the old IWW. Finally, on July 20, the old Federation of Woodworkers convened as the new International Woodworkers of America — the IWA — an autonomous, industry-wide, CIO lumber workers union.

The Tacoma local with its three thousand members refused to support the move to CIO affiliation. Homer Haney, business agent for the local, reiterated the "railroading" charges and repeatedly questioned the propriety of affiliating so quickly. On the last day of the convention the whole Puget Sound District Council withdrew from the IWA and announced that it would remain in the AFL to do combat with the new lumber workers' union.[24]

The convention remained overwhelmingly pro-CIO and elected Harold Pritchett to be the first president of the International Woodworkers of America. Pritchett's election was bitterly opposed by leaders Al Hartung and Don Helmick from the Columbia River District Council in Portland. It was no secret that the two men coveted the leadership of the lumber workers' movement for themselves, and now that the prize was an international presidency, they were all the more strident in their opposition to Pritchett. Hoping to nullify his election, Hartung and Helmick proposed that it should be ratified by a rank-and-file referendum with the results to be disclosed at the December, 1937 IWA convention. This proposal was accepted by the convention, but to the dismay of Pritchett's opponents, the referendum ratified his election.[25]

Jurisdictional Warfare

Such a large defection of lumber workers to the new CIO industrial union movement could not go unchallenged by the AFL which had contributed organizers and financial support to the lumber workers. Finding themselves unable to woo back the disobedient lumber workers, the Carpenters set out to crush the rebel IWA. At the 1936 UBCJ convention, General Secretary

Frank Duffy had promised the Federation of Woodworkers the "sweetest fight" of their lives if they left the Carpenters Brotherhood.[26] And he threatened to place on the AFL's "Unfair List" all firms that continued to employ lumber workers who were outside the United Brotherhood. In Portland, a year later, Charles Campbell, secretary of the AFL Building Trades Council, put teeth into Duffy's threat by announcing: "We refuse to erect any lumber manufactured under the CIO banner. The Carpenters don't consider the CIO a democratic organization."[27] On August 11, 1937, the United Brotherhood issued a circular letter that set the tone for their campaign against CIO lumber workers:

> The Committee [of the general executive board] found that there were Communist and adverse influences boring from within for the purpose of trying to destroy the activities of the United Brotherhood...To combat this dual movement...our members must not handle any lumber or mill work manufactured by any operator who employs CIO [lumber workers]...Let your watchword be "no CIO lumber or mill work in your district," and let them know you mean it.[28]

After this, Teamsters president Dave Beck threw his weight behind the Carpenters.

When the International Longshoremen's Association and Harry Bridges decided to back the IWA, the AFL leadership made their position explicit: the United Brotherhood would boycott all CIO-produced lumber and mill work nationwide and instruct locals throughout the country not to handle CIO-produced materials. If lumber came from CIA mills that had not been closed by AFL pickets and the longshoremen loaded it aboard ships, the AFL Teamsters would refuse to move it from the docks to the job sites. If some lumber did manage to arrive at a job site, the AFL building trades craftsmen would refuse to handle it.[29]

In preparation for the impending boycott and jurisdictional dispute with the International Woodworkers of America, the United Brotherhood of Carpenters and Joiners gathered their still faithful representatives together at Longview, Washington on August 21, 1937 to form them into the Oregon-Washington Council of Lumber and Sawmill Workers. Many of the AFL leaders believed that the boycott in and around Portland would be decisive, for the Carpenters were well organized in that city. An AFL victory there would greatly limit the success of the CIO campaign to organize lumber workers throughout the rest of the Pacific Northwest.[30]

The AFL acted swiftly. On August 16, the Portland Central Labor Council voted 175 to 69 to suspend the 4,000 lumber workers who had left the AFL Sawmill and Timber Workers Local 2532 for the new IWA Local 3 of the CRDC. Since no AFL lumber workers remained in Portland, the Building

Trades Council was called on to picket mills where lumber was produced by the CIO members. Seven of the city's ten major operations were forced to shut down, leaving twenty-five hundred men out of work.[31]. The boycott was further strengthened when the Japanese invasion of China restricted the shipment of mill products to Asia, forcing still more IWA mills to close down or operate part-time. With the mills closed, the AFL had effectively blocked the IWA's claim to representation.[32]

The IWA and its president Harold Pritchett sought a settlement of the jurisdictional dispute through the National Labor Relations Board.[33] Regional NLRB director Charles Hope tried in vain to get the AFL to agree to a representational election, but the AFL steadfastly refused to recognize the Board's decision to certify the IWA as collective bargaining agent in seven of Portland's twelve lumber mills, where sixty-eight to ninety-two percent of the work force was in favour of the IWA.[34] Instead, to both bolster their picket lines and enforce the boycott of CIO-produced lumber, AFL "goons" from the Teamsters local overturned trucks loaded with lumber from mills where the AFL contested the CIO certification, and beat up the workers.[35] Sixty miles south at Salem in the Willamette Valley, Al Rosser, secretary of the Teamsters local, set fire to a box factory where employees had voted for the CIO. Convicted of arson in 1938 and sentenced to twelve years in prison, he was paroled from the state penitentiary for good behavior in 1943 although the Oregon Supreme Court had unanimously upheld his conviction.[36]

In short, "all hell had broken loose" around Portland during the fall of 1937. If the AFL could just stop the IWA drivers from delivering fuel wood and sawdust to city homes, the remaining mills would have to close as the stuff was too bulky to stockpile for long. Again and again CIO drivers were beaten, but by the time police arrived, the goons had fled. And if the CIO men armed themselves with clubs, they were generally arrested for carrying concealed weapons. Meetings held with Mayor Carson, the Portland City Council, the governor and the police to stop the attacks were to no avail. But B.A. Green, attorney for the IWA, had a novel suggestion: why not carry softball bats, a glove or two and some softballs in the trucks? When the goons showed up, the bats could be unlimbered; when the police came, the CIO was only holding a softball game! Two weeks later, the goons had disappeared.[37]

The [Portland] *Oregonian*, although never favorable to the CIO, blamed the AFL for most of the fights and property damage that accompanied the jurisdictional struggle. The September 3, 1937 issue editorialized that:

...the workers in the mills are CIO men...the AF of L pickets and beat-up thugs who have been following the CIO workers about and assaulting them are in the role of interlopers. They are doing their best to terrorize workmen who are union men like themselves and whose only offense is that they persist in working [at the CIO mills]. They openly threaten mass picketing, with the increased violence certain to go with it. By these means they are about to shut down Portland's greatest industry. [38]

For the IWA, support from the longshoremen boosted morale. Dockers spoke in favor of the CIO union and contributed financially at a time when the IWA had few funds of its own to support the workers laid off by the boycott. [39] But the IWA-ILA relationship also had its contradictions. When Portland's docks were picketed by IWA members, Portland longshoremen refused to load AFL lumber from the Carnation Mill in Forest Grove, Oregon. But when the Waterfront Employers Association threatened to close down the entire port unless the lumber was loaded, Harry Bridges asked Pritchett to remove the pickets. Pritchett did so and thus angered Al Hartung of the CRDC. At the first IWA convention in December of 1937, Pritchett's concession to Bridges would become a wedge which the CRDC hoped to drive between the lumber workers and the IWA leadership. [40]

The AFL repeatedly refused to meet with the NRLB representatives or accept offers of conciliation; they were determined to break the IWA and would not forsake their boycott. Finally, Oregon's Governor Charles Martin had had enough. He rebuked both Charles Hope and the NLRB, and announced that he would conduct his own election at Portland's large Inman-Poulsen Lumber Company. In the balloting which followed the IWA defeated the Carpenter's local 376 to 183, but the AFL simply disregarded the results. [41]. Martin then denounced the AFL and moved to end the shut-down. When the mills reopened to their anxious and financially depleted workers, state police were standing by with orders from Martin:

... Stand off at a respectable distance and read the Riot Act in a loud, clear voice. Then order them [the AFL pickets] to disperse. If they don't, go into 'em and beat hell out of'em. [42]

The majority of Portland's mills had been shut down for four months, but when operators began to reopen the mills in late December, 1938, some posted "company union" work conditions, others favored the AFL. The IWA was forced to continue its battle through the auspices of the NLRB, charging that lumber operators had again either violated IWA certification and locked the union out, or refused to bargain in good faith. It was clear to the IWA executive officers that AFL leaders and many of the employers were in cahoots. Companies were still recognizing AFL mill charters in total disregard of NLRB rulings. [43] The AFL accommodated the mill owners by calling off the boycott and their pickets, making it possible for the plants to resume operations.

The legal struggle of hearings and votes continued until 1939. In the end, it was not a clear-cut victory for the IWA, but neither had the AFL been able to crush the rebels. The IWA-CIO controlled six Portland mills and the AFL five, but total membership tilted two to one to the IWA because it controlled all but one of the larger mills. The IWA's success made its left-wing leaders the immediate targets of local and federal investigations into alien radicalism. Soon, the terrain of battle would shift from the picket line to the realm of political ideology.

The Portland Red Squad and Dies Committee Investigations of Alien Radicalism

While there were fertile conditions for anti-communism in the Pacific Northwest, opposition to the IWA's leftist leadership was greatly stimulated by the climate of fear established by the search for "reds" conducted by the Portland authorities. The Portland Police Red Squad, originally formed in 1920 to help break up the IWW, continued to function throughout the 1920s and 1930s, but by 1937, Walter B. Odale, who had worked with the Red Squad since its inception, was the only remaining active member.[44] The organization was supported and approved by Odale's superior officer, Detective Captain John J. Keegan, and by Portland's mayor, Joseph B. Carson, who assigned city policemen to assist with Red Squad investigations. By 1937 Odale had collected the names of 10,000 alleged Communists and Communist sympathizers who were involved in Communist-supported or -sponsored activity in the Portland area. At the same time, he calculated that there were only 132 actual Communist Party members in Portland.[45]

Odale's primary concern was to discredit the industrial union movement.[46] His investigations delved primarily into the Federation of Woodworkers, the International Woodworkers of America and the Portland Industrial Union Council, and secondarily into such organizations as the Oregon Commonwealth Federation and the local chapter of the North American Committee to Aid Spanish Democracy. According to Odale, "...the entire Federation of Woodworkers [and the CIO] movement for industrial unionism is best understood as an expression of the communist movement in the West."[47]

Between February 19 and October 19, 1937, Odale issued a series of five-to-six-page mimeographed reports on his investigations which he entitled *Weekly Report of Communist Activities*. Although most of his information was gleaned from local and national Communist pamphlets which had been distributed on the streets in Portland, Odale labeled all his reports "Strictly Confidential: Do Not Make Public." The *Reports* were nevertheless made available to employers on request, and copies were regularly sent to Mayor Carson and Oregon Governor Martin.[48]

Americans Incorporated, an organization dedicated to "Fostering American Ideals and Perpetuating Our [American] Institutions," irregularly issued a two- or three-page *Radical Activities Bulletin* to expose the alleged subversion within the industrial labor movement.[49] Their investigations centered on the union activities of Harold Pritchett, and they blamed him for having created the turmoil over industrial unionism in the Northwest. On February 14, 1939, Americans Incorporated sponsored a "National Defense and Americanism Rally" to discuss "whether the IWA president is the type of alien Communist that we [Americans Incorporated] want at the head of our labor organizations."[50] Charles H. Martin, now the former governor, was rally chairman, Portland Mayor Joseph Carson was master of ceremonies, and Portland Police Captain John J. Keegan was program director.

The 1935-1937 American Legion Subversive Activities Committee was led by George Stroup, who later became secretary of Americans Incorporated. As representatives of the American Legion, Stroup and his agents entered mill towns and persuaded employers — usually mill owners — that an investigation of local subversive activities would help control or eliminate the local movement for industrial unionism by silencing or exposing the radical labor organizers. Oregon employers contributed approximately $11,000 to support the investigations; in return they received a secret report containing lists of alleged Communists or left-wing labor organizers.[51] Many mill workers' names appeared on the Legion's subversive list even though they had never participated in the industrial labor movement and were not connected with the Communist Party. A worker could only get his name removed from the list and protect his job by going to the offices maintained by Stroup or the Portland Police Red Squad.[52] There — after much delay and trouble — he could obtain a statement correcting the error.

After public disclosure of its tactics in 1937, unfavorable publicity forced the Legion to dissolve the Subversive Activities Committee, and thereafter, the Police Red Squad investigations were directed by Detective Captain Keegan. Odale no longer operated independently; his weekly *Reports* ceased and his card files listing thousands of names were moved into the Portland Police headquarters.[53] It seems very likely that the techniques which Odale had been using were extra-legal if the methods used in the Portland Police

Department's investigations after 1937 offer any criteria for judgment. On March 16, 1938, W.A. Lewis, a Portland Police officer, summarized a raid on the Commonwealth Cafe, known to be frequented by union organizers: "A customer, Harold Spring, who was suspected of being a left-wing CIO organizer, was arrested and taken to the police station where he was questioned, searched and relieved of his personal papers." Another report dated February 17, 1938 was prefaced by the comment: "The following information is confidential, being given to us by an informant who took the young lady in question, Miss Frances Hallander, out socially for the purpose of obtaining this information."[54]

The investigations of alleged subversive activity frequently resulted in random indictments being directed at organizers within the industrial labor movement, but Harold Pritchett, president of the International Woodworkers of America, and Harry Bridges, president of the International Longshoremen's Association and West Coast CIO Director, were selected for more intensive investigations. The close working relations with Communists which both Bridges and Pritchett acknowledged made them suspect to certain employers and to the American Federation of Labor. Both men were frequently labeled "alien labor agitators." Bridges had emigrated from Australia in 1920 and still lacked his citizenship papers; Pritchett was a British-born Canadian citizen, admitted into the United States on a visitor's permit. Pritchett had first entered the United States in 1936 as a delegate to the first convention of the Federation of Woodworkers and was later elected to the presidency of that organization. At that time, the Central Labor Council AFL had unanimously adopted resolutions calling for a permanent visa for Harold Pritchett. But by late 1937, after Pritchett had successfully organized the lumber and sawmill workers into a rival CIO union, the Council reversed its position and passed resolutions labeling Pritchett a radical and a subversive labor oganizer, and demanded his immediate deportation to Canada.[55] Council secretary Gust Anderson urged Secretary of State Cordell Hull to conduct a full investigation into Pritchett's possible subversive activities.[56]

During a 1937 National Labor Relations Board Inquiry into the jurisdictional disputes between the International Woodworkers of America and the United Brotherhood of Carpenters and Joiners, the AFL's attorney accused Pritchett of being a Communist. As evidence, he submitted a photostatic copy of an obituary printed in the October 9, 1935 issue of the *Surrey Leader*, a Cloverdale, British Columbia, weekly. The obituary said in part:

On arrival of the cortege at the Curry Centre Cemetery, the leaders of the Communist Party took charge of the remains. A short address was delivered by Comrade Pritchett of Coquitlam, the Red Internationale was sung.[57]

During the court recess, Pritchett readily admitted speaking at the funeral and said he would ". . . go to the funeral of a friend, Communist or no Communist," regardless of how the *Surrey Leader* referred to him in an obituary.[58] The column was soon touted by lumber operators and AFL officials as solid evidence that Pritchett was a Communist. Indeed, this was the only piece of evidence against Pritchett that immigration authorities or police departments ever made public.

However, after Pritchett was elected president of the International Woodworkers of America (ratified by referendum in December 1937), his visa was withdrawn for the stated reason that he was no longer admissable into the United States under the laws governing the entrance of foreigners. After that, he continued to travel to Portland for IWA conventions and for Columbia River District Council conferences on a visitor's permit.

Now it was the turn of the Portland Central Labor Council to assail Pritchett for creating unrest in the lumber industry: "the market for lumber has been diverted to the Canadian lumber operators by the activities of Harold Pritchett, to be manufactured by Oriental labor, and at cheap wages. . ."[59] One of Oregon's largest lumber exporters, Dant and Russell Incorporated, blamed both Pritchett and Bridges for mill closures and urged Congressman Walter M. Pierce to have them deported:

Most of the unemployment in our lumber industry can be traced directly to the subversive activities of two foreign radicals, one a native of Canada and one a native of Australia, who have no respect for the laws of this country or the citizens. Immediate deportation of these two men would do more to establish normalcy in our lumber and shipping industry than any other one thing that can be done.[60]

In August 1938 the House Committee on Un-American Activities, chaired by Martin Dies of Texas, launched an investigation of Pritchett. As usual before such hearings, impartiality was not a prerequisite for witnesses; none of them disguised their dislike of the CIO union movement. John P. Frey, president of the Metal Trades Department of the AFL, testified that ". . . the president of this organization [the IWA] is Harold Pritchett, a Canadian Communist."[61] Harper Knowles, a self-appointed California expert on radicalism who had testified during the deportation hearing for Harry Bridges, also appeared to give testimony against Pritchett. His official title was Chairman of the Radical Research Committee of the American Legion, Department of California. He had also worked as a private detective for various West Coast police agencies investigating alleged union subversives. (His "expertise" and bias would later land him a job with the Associated Farmers of California, an organization not known for its sympathy towards farm labor unionization.) Knowles stated that Pritchett was ". . . an active member of high standing in the Communist Party," and that he pursued a program of ". . . disrupting the lumber industry in this country."[62]

In further testimony before the House Committee, Captain Keegan of the Portland Police Department's Red Squad insisted that Pritchett was a Canadian Communist who "was down [in] Oregon during the last election campaign trying to advise the people how to vote." To Keegan, both Pritchett and Harry Bridges merely "worked under the guise of labor leaders but in reality they were Communists." Everywhere he looked the Detective Captain found that union organization was secondary to subversion; Bridges' and Pritchett's "real purpose was to overthrow the U.S. government by force and violence."[63]

The Effect of the Red Scare within the IWA

The first regular convention of the International Woodworkers of America opened in Portland on December 3, 1937. Press representatives were barred from the meeting, but Al Hartung and his supporters from the Columbia River District Council parlayed the presence of a *Daily Worker* correspondent into a confrontation on the issue of Communist lumber workers in the union. They warned against the ". . . stigma that will be placed on the entire international union by having the Communist Press seated in the convention." After one and a half hours of discussion, a motion to exclude the correspondent failed, 136 to 106.[64]

Later during the convention, the Hartung faction charged that the IWA administration and Harry Bridges had collaborated to ". . . dictate strike policy to the CRDC" when it had struck the Carnation Lumber Company in Forest Grove, Oregon on November 3, 1937.[65] The CIO had won the NLRB election to represent the Carnation employees, but the management had locked out the IWA. In retaliation, the CRDC Local 3 had established picket lines in Portland shutting down the docks where Carnation lumber was being loaded aboard ships. Harry Bridges, fearing that the Waterfront Employers Association would declare a coast-wide lockout, persuaded Pritchett to advise the CRDC to withdraw the picket lines from the waterfront.[66]

Delegates from the CRDC testified that the IWA executive board needlessly adhered to Bridges' instructions during the strike. The IWA administration, they contended, was no longer devoted to improving wages, hours and working conditions. In reply, Pritchett maintained that the picket lines would have affected public opinion adversely and would have forced longshoremen and lumber workers into a coast-wide strike which neither union could afford.[67]

Dissatisfaction over the Carnation Mill strike developed into a bid to take control of the union. The small district council from Coos Bay joined with the CRDC and nominated opposition members Al Hartung for president and Kessler Woodruff for executive secretary. Hartung, a native-born American who had been a logging foreman in Oregon, was a dedicated anti-communist. Pritchett's bid for renomination received the support of seven district councils. In the referendum that followed the convention, Pritchett, Vice-President O.M. Orton, and Executive Secretary Bertal McCarty – all incumbents – were reelected.[68]

The factional strife resumed at the second IWA convention in Seattle, September 12 to 16, 1938. Since the first convention, the IWA local in Portland had overcome the employers' nine-month lockout that had developed after affiliation with the CIO, but the local was still plagued by a jurisdictional dispute with the American Federation of Labor. Throughout this dispute, the AFL's Central Labor Council and the United Brotherhood of Carpenters and Joiners had relied on the investigative reports provided by the Red Squad, Americans Incorporated, and the American Legion to label the CIO's Portland Industrial Union Council and the IWA "Communist dominated."

Prompted in part by these reports, the Portland IWA Local 3 introduced a resolution to amend the union's constitution to state that its membership requirements did not conflict with the U.S. Constitution. The CRDC assumed this would help stifle criticism by the locals that the IWA granted union membership to subversives.[69] A second resolution, which would have barred Communist Party members from office, posed an even more direct threat to the administration. A third resolution was introduced to censure IWA vice-president Mickey Orton for his intervention during the Carnation Mill strike. All three resolutions were defeated. The Columbia River District Council had confronted the administration for a second straight time in convention and had failed. Angered, Al Hartung threatened to withdraw the CRDC from the IWA if its resolutions were not granted greater consideration in the future.[70] But it was clear by this time that Hartung's anti-communist faction was a minority, and unless the political balance of the IWA could somehow be shifted, the union's political course would not be altered.

In the Spring of 1939 a group led by former leftist Harry Tucker defeated the Communist-backed incumbents in the leadership race in the southwestern Washington Aberdeen local. The origins of the rift in the local can be traced to the same historical schisms that had influenced the Raymond local in 1935. In Aberdeen, a faction emerged similar to the one which had taken a conciliatory line in Raymond's anti-Bedaux struggle, but unlike Raymond, the collaboration of the Aberdeen rightists extended beyond the politics of in-plant issues and fed into the general anti-Communist movement that was developing in the Aberdeen community.

Behind this movement was the Aberdeen Better Business Builders, formed in January 1939, to counter the effects of the "Roosevelt recession." Although the stated purpose of the organization was the enhancement of cooperation between management and labor, membership in the organization was restricted to businessmen, farmers and professionals. Russell V.Mack, an American Legionnaire and publisher of the *Grays Harbor Washingtonian,* and Joe Clark, editor and publisher of Aberdeen's *Grays Harbor Post* were on the Better Business Builder's executive board. Their papers gave the organization much favorable coverage. The only unionist who was known to work with the group was Harry Tucker.

The main function of the Better Business Builders was to provide pro-employer and anti-Communist propaganda. Through public meetings and handbills the organization campaigned for federal aid to the lumber industry and for the deportation of Harold Pritchett and Harry Bridges. On June 6, 1939 the Better Business Builders put on a banquet and night of entertainment — "a strictly masculine feed" — for over two thousand AFL and CIO union members at the Aberdeen Elks Club. In his speech to the gathering, Russell Mack presented the Better Business Builders as a friend of labor; the real enemies of both labor and capital, he told them, were the Communists who unfairly maligned the employer group. It was evident from Mack's speech that his purpose was not labor-management unity so much as an occasion to sow dissension in the ranks of the IWA.[71]

For workers on Grays Harbor who had suffered several years of unemployment and numerous violent strikes, Mack's promise to cooperate with non-Communist unionists like Harry Tucker was no doubt encouraging. Moreover, the general anti-Communist climate being whipped up by the Better Business Builders promised nothing but more repression for those who remained sympathetic to the Left. It was against this backdrop and with the Better Business Builders claiming partial credit that Harry Tucker and the anti-Communist bloc in Aberdeen won election in 1939.

Following the election, Dick Law, leader of the Communist faction, charged that Tucker and other anti-Communists were in collusion with the business community. The International newspaper, *Timberworker,* backed Law and cited as evidence the collaboration of Tucker and the Better Business Builders of Aberdeen to extort a political defection from William Anderson, secretary of the Grays-Willapa Harbor, Washington IWA District Council. In a story on June 3 — "Expose the IWA Wreckers" — the paper told how Anderson, who was not an American citizen, had travelled to Vancouver, B.C. in January, 1939 and been detained at the border when attempting to re-enter the United States. Five months later when Anderson was still awaiting permission to re-enter, members of Tucker's faction, Joe Clark, James Fadling and John Deskins, went to Vancouver to offer him a deal: if he would swear that he was not a Communist and if he would defame the left-wing leaders of the IWA, the right wing leaders of Aberdeen would lean on the immigration authorities to grant him a visa. Unbeknownst to the plotters, the meeting between them and Anderson was being recorded. A bugged flower pot wired to a recording machine operated by Ernie Dalskog, a left-wing IWA member from British Columbia, recorded the whole conversation. The meeting's proceedings were reported in the *Timberworker* and the tapes were played over the radio.[72]

Tucker's faction, however, tried to turn the tables by charging that the incident was a frame-up calculated to discredit their effort to defeat Dick Law and the IWA's administration. A committee was established to investigate the allegations made in the *Timberworker;* after an extended review, the committee found the evidence insufficient to support the charge that the Aberdeen local had purposely collaborated with the Better Business Builders.[73]

The Third Convention of the IWA was to open on October 18, 1939 in Klamath Falls, Oregon. That afternoon the union's executive board met in special session. John Deskins, Harry Tucker and Joe Clark burst unannounced into the meeting and charged international officers Harold Pritchett, O.M. Orton and Bertal McCarty, plus executive board member Dick Law, with having violated the union constitution by using the *Timberworker* to expose the Anderson incident. Executive board member Don Helmick joined in the attack, demanding that the international officers withdraw from the meeting and allow the remaining executive board members to try them. The officers refused to leave and the confrontation simmered. Finally, Helmick, the Aberdeen members, and three other board members walked out.[74]

That evening following the opening session, delegates from the CRDC, the Boommen and Rafters district council, and participants from the Tacoma, Aberdeen and Raymond locals held a caucus at the local IWA hall to discuss

opposition strategy for the convention. Don Helmick vehemently maintained that the administration must be defeated because the "...policy of
the IWA is laid down and promoted by instruments [the Communist Party]
outside the lumber industry." In a public statement, caucus delegates
"pledged themselves to free the organization of all communistic influence
and from the attempted domination of communistic sympathizers."[75]

The following day, Robert Williams, an administration supporter from
Longview, divulged Helmick's statements to the entire convention.
Williams asserted that Helmick had accused IWA leaders of lying and conducting dishonest referendums, and that the caucus had laid plans to put
Helmick in control of a move to oust the IWA's leadership. A madhouse
ensued. Boos and jeers reverberated across the convention floor. Helmick
jumped to his feet to answer Williams, declaring that the CRDC was "defi-
nitely opposed to the present administration of the IWA" because the executive officers had mismanaged the union and because Communists, namely
James Murphy in Portland and Morris Rapport, Communist Party leader
in Seattle, had formulated the organizing policy for the IWA. Helmick
accused Harold Pritchett of encouraging him to join the Communist Party
in February 1937, in order to run for office in the old Federation of Woodworkers. "If I refused," he told the convention, "he said he would run me
out of the industry." Helmick insisted that the official IWA membership
figure of 100,000 had been exaggerated by the administration to gain
support, and that the actual figure was approximately 25,000.[76]

Helmick recalled the 1937 Columbia River District Council strike against
the Carnation Lumber Company at Forest Grove. He asserted that the
picket lines around the Portland docks had been necessary and that Harry
Bridges and Pritchett had "sold out" the Council by refusing to support the
decision to picket the docks. Replying to this charge, Harry Bridges, who
was attending the convention, reiterated the position of the ILWU as established at the first IWA convention: that the longshoremen were not pledged
to support every "wildcat" strike to the finish and would not endanger the
entire longshoremen's union or the IWA for a strike called solely by the
Columbia River District Council.[77]

When it was his turn to address the convention, Pritchett submitted an
affidavit that stated he had never been a member of the Communist Party.
He denied that he had asked Don Helmick to join the Communist Party in
February, 1937: "...I couldn't if I wanted to. I am not a member of the
Communist Party. Besides," he said, "I don't think the CP would accept
Brother Helmick."[78]

A heated discussion produced two resolutions which were intended to
resolve the division between the IWA administration and the opposition.

Harold Pritchett proposed that a Unity Committee meet in conference and develop a policy to reconcile the opposing forces for the coming year. His proposal passed by 123 to 107 on a roll-call vote despite the unanimous opposition of the CRDC. The Unity Committee included the IWA executive board, the officers of the CRDC, and the regional CIO directors for California, Oregon and Washington: Harry Bridges, William Dalrymple and Richard Francis. A meeting of these participants failed because they were unable to agree on the distribution and tabulation of votes within the Committee.[79]

The second proposal directed the three regional CIO directors to submit a unity plan to the convention. They held a special hearing to give spokesmen from various district councils an opportunity to voice their complaints about the union and its officers as well as to offer proposals to resolve those complaints. Their report diplomatically side-stepped any direct reference to Communist Party members or their sympathizers. It simply reiterated that the national CIO granted autonomy to every union in determining its membership, but because the IWA's official newspaper, the *Timberworker*, had frequently carried partisan editorials and articles that criticized Pritchett's opposition, the report recommended that in future the *Timberworker* should be entirely devoted to the problems of the lumber industry, that it refrain from criticizing any IWA member, that it give all IWA candidates equal news coverage, and that it reflect the policies adopted at the IWA and CIO conventions. The regional directors anticipated that these recommendations would end the dissension created by the editorializing in the *Timberworker*.[80]

But the regional directors' report failed to satisfy the Columbia River District Council which was determined to exclude Communist Party members from the union. Quickly, CRDC delegates moved to amend the IWA constitution, to wit:

> No worker, otherwise eligible to membership in the union, shall be descriminated against or denied membership by reason of race, color, religion, or political affiliation, except membership in, or the support of, Communist, Nazi, or Fascist movements or subsidiaries thereof.[81]

Delegates testified that Communist Party members had disrupted IWA locals and interfered in union organizing activities by issuing broadsides and handbills that supported the union. Lumber workers who might otherwise have joined the union were allegedly deterred by these Communist Party activities. Harry Bridges condemned the incessant factionalism and urged that delegates defeat the amendment. No one, he declared, had established that Communists were, in fact, destroying the IWA, and a prolonged debate over who should or should not be an IWA member would only bleed the union and provide grist for AFL propaganda. The administration survived the challenge 123 to 99 with the CRDC casting 49 votes for the amendment.[82]

This was a major defeat for the IWA's anti-communist faction, the AFL, and the business community who were united in their opposition to the Pritchett leadership. The referendum elections that were to be held in five weeks were their only remaining hope for changing the union's course. The campaign that they launched to take over the union pitted an opposition slate headed by Al Hartung and Harry Tucker against the Pritchett administration. Then on December 1, three days before the voting period ended, red scare tactics were renewed in Aberdeen, and within a month anti-communist terror had immobilized the community.

There was to be a dance at the hall of the Finnish Workers Federation in Aberdeen the night of December 2, 1939. This hall was the property of the so-called "Red Finns" because most of the Finns who belonged to the Federation were either Communists or socialists, and since the days when the IWW had met there, their hall had been known as the "Red" hall. With the Russo-Finnish War underway, the Red Finns had become increasingly isolated from the "White Finns" in the community, who were staunchly anti-soviet.[83]

Either by design or oversight, the White Finns announced a relief rally to raise funds for Finland's defence to be held on December 2, the same date as the previously announced Red Finn dance. The non-Finnish elements in the community seized the opportunity to organize a picket of the dance at the Red Finn hall. White Finns disclaimed responsibility for the planned picket, although they had asked the inflammatory Russell Mack, the well-known anti-communist, to speak at their rally.

The night before the Relief Rally, the Grays Harbor Central Labor Council adopted a resolution which condemned both the Soviet Union's invasion of Finland and the Red Finn dance. A further demand was tacked on to the resolution: Harold Pritchett should be deported. Thus, for many Harbor zealots, a picket of the Red Finn Hall would be an attack on the IWA radicals and the Communists, even though the Finnish Workers Federation had never endorsed Communism *per se*.

The day of the dance and the rally, the Aberdeen *World* revealed who was behind the planned picketing:

AFL union members and World War veterans this afternoon were said to be planning to picket the Grays Harbor community 'victory dance' tonight.... Spokesmen said citizens became aroused because Communists placed their dance posters over those announcing the Finnish rally in the Randall Street hall to launch a drive for funds to aid stricken war victims in Finland.

Those planning to picket the Workers Hall will gather in front of the Oregon-Washington district A.F.L. sawmill union hall on Wishkah Street at 7:45 o'clock, it was said.

A leader will be chosen and the march will begin with the American flag at the head. The "pickets" will march through town to the Workers Hall. [84]

To avoid the threatened violence, the Red Finns cancelled their dance, and the protest parade from the AFL headquarters failed to materialize. Nevertheless, three or four hundred demonstrators marched on the darkened Finnish Workers Hall, smashed in the door and ransacked the building. Furniture, floor boards, stage scenery and a piano were all reduced to kindling, and water flooded the floor from a broken fountain. Books and pictures of Joseph Stalin and Earl Browder were consumed in a street bonfire.

Heinie Huff, Grays Harbor Communist Party executive secretary, fixed the blame for the Red Hall destruction on the "witch-hunting campaign of Martin Dies, the Better Business Builders and the local press, and on the laxity of the law enforcement officers." Left-wing forces in the CIO called for a thorough public investigation. The police finally arrested two suspects, Ward Penning and Joe La Londe, an executive board member of the AFL's Oregon-Washington Council of Sawmill Workers. Both pleaded innocent, and both claimed the Communist Party was out to defame them because they had previously refused to support Communist union activities on the Harbor. A jury trial found them not guilty after twenty minutes of deliberation. [85]

What effect the wrecking of the Red Finn Hall had on the election outcome is unknown. Harold Pritchett, O.M. Orton and Bertal McCarty were reelected, but a constitutional amendment adding a second international vice-presidency also passed and Worth Lowery from the Columbia River District Council was elected to that post. Lowery's election represented the first breach in the left wing control of the International. The significance of his election, however, was momentarily overshadowed by the vigilante activity that had marked the election period and the fear that the wrecking of the Finn Hall was just the beginning.

Who Killed Laura Law?

On January 5, 1940, just five weeks after the Red Finn hall had been ransacked, Laura Law, wife of Dick Law, Grays Harbor's best-known and most militant IWA leftist, was murdered. Dick Law had been in downtown Aberdeen at a union meeting. Laura Law's parents, who lived in the same house, had discovered the body when they returned home from a movie. [86]

Shivers ran through the families of radical unionists, for the slaying had been especially brutal: seven 6½-inch stab wounds in the breast; five blunt blows to the head that smashed and contorted her mouth, plus two blows upward across the right cheek which fractured the skull; arms, legs, and body bruised. Newspapers and magazines were scattered about the living room and dresser drawer contents were spilled on the floor.

Dick Law was originally from Oregon. At fifteen he had been rigging in logging camps as the sole supporter of his mother, brothers and sisters. Then, when the family needed money for hospital bills, Law had broken into a grocery store in Tillamook, Oregon. Arrested and convicted, he served a short sentence in the Oregon State Penitentiary. In 1931 he was arrested again in Klamath Falls for assault with a deadly weapon, but the charges were dropped and Law was commended by the judge when he explained that the accident occurred when he was being robbed and that he had fought back. After he was blacklisted for his union activities, he had drifted from camp to camp in Oregon and settled in Grays Harbor in 1934. By 1935, the twenty-eight year old Law was an established union leader in the forefront of the Northwest Joint Strike Committee's rebellion against Abe Muir. In the next five years he developed wide support among left-wing IWA members because his energy and strike activities had noticeably improved wages and conditions for saw millers and loggers in the Grays Harbor area.

In 1935, Law married Laura Luoma, who became a solid unionist in her own right. During the 1935 strike, she helped maintain a soup kitchen for three to four thousand strikers. In the following years she organized women's auxiliary branches of the IWA and in 1937 was elected the first president of the Aberdeen Women's Auxiliary. That Laura Law had many friends in the IWA-CIO movement was clear when three thousand loggers and mill workers gathered for her funeral.

But the couple had paid a price for their militancy. It was common for them to receive threats from anti-communists, from lumber workers who favored the AFL and from lumbermen concerned with narrowed profit margins. Law requested, but never received, police protection after the Finn Hall episode, which he recognized as a sure sign that the vigilantes would now stop at nothing to slow the industrial union movement.

A pamphlet distributed in 1937 had called for Law's "liquidation," and in 1938 his wife had received numerous threatening telephone calls. Businessmen's organizations, such as the Committee of Industrial Stabilization and the Grays Harbor Citizens Committee, had joined the anti-Law barrage, issuing broadsides which called for the removal of Law, Bridges and

Pritchett. The Order of Better Americans, reportedly an AFL Lumber and Sawmill Workers front group, joined the fray. One businessman in particular was reported to have made numerous threats against Law; "one way or another," he is said to have stated, Dick Law would have to go. And to top it off, *Crow's Pacific Coast Lumber Digest* issued a regular litany of articles and editorials that couldn't help but agitate the anti-radical businessmen's factions in Aberdeen: "As we stand today a mere handful of federal-backed reds are bulldozing the 8,195,000 people of the Pacific Coast into bankruptcy. Oregon, Washington and California have never had a better cause. What are we waiting for?" Anyone who was receptive could find these messages provocative enough to warrant taking action against Dick Law.[87]

On January 14, an agonizingly slow nine days after the murder, the first inquest convened. Dr. John Stevenson, Grays Harbor coroner, headed the hearing. On one side sat Paul O. Manley, deputy Grays Harbor prosecuting attorney; appearing for Dick Law and his family were John Caughlan, the Seattle executive secretary of the American Civil Liberties Union, and Irwin Goodman, a Portland labor lawyer. Stevenson appointed six jurors: two insurance dealers, a businessman, a barber, a service station manager and a stevedore. Goodman protested that this was not a balanced selection, but his request to question the members' qualifications was brushed aside by the coroner. This was an inquest, not a trial. But to radicals in Grays Harbor it already looked like an attempt was going to be made to becloud the issue of vigilante activity by casting suspicion on Law himself for the murder of his wife.

Where was Law the night of the murder? "Downtown in the block where [my] office [is] located." No contrary testimony as to Dick Law's whereabouts was ever offered. What about Law's prison record? Was the house ransacked by someone to obliterate the real motive for the murder?

"You are not a Communist, are you, Mr. Law?"

"No."

"Do you have any enemies in the communist ranks?"

"I don't know. I don't know who [all] the Communists are. I know a few Communists."

"Have you ever had any difficulties with any of them?"

"No."

Law admitted to having copies of the Constitution and by-laws of the Communist Party in his house. They had been given to him since he was in the habit of studying material on the Far Left as well as the Right. And lastly, what about Law's married life? Was he infatuated (or involved) enough with union secretary Helen Soboleski to ask his wife for a divorce? And had Laura Law refused? Law certainly had seen Soboleski socially, but denied any discussion of divorce. The fact that Law married Soboleski some years later kept the issue of their relationship prior to the murder alive, but in the context of the coroner's inquest, the issue of the relationship betwen Law and Soboleski seems to have been little more than an attempt to impugn Law's integrity.

Dick Law's attorneys took several tacks. They argued that the long record of threats against Law and his wife could not be discounted as motives for murder. Those who wanted to destroy the IWA industrial labor movement certainly would not hesitate to attempt to intimidate Law by murdering his wife. But the prosecution would not concede that the threats and the murder were related.

> Law: *(to Prosecutor Manley)* If you are serious...you will
> know that the bombings, the wrecking of the Finn Hall,
> this campaign of hatred and the murder of my wife are
> inseparable and cannot be divided.

> Manley: I don't know that.

> Law: You mean you don't want to know that.

The tie to the wreckage of the Finnish Workers Hall was important to Law. He had compiled a brief on the event that implicated several local citizens. Had his wife been tortured to force her to divulge its location? Law believed this was central to the case and that it would explain why the house had been partially ransacked.

Then Law dropped a veritable bomb in the courtroom: would the prosecutor subpoena the people Law had mentioned to him as possible suspects?

> Manley: Will you name the people now?

> Law: I will be glad to if you honestly want them.

> Prosecutor: Name them, go ahead.

> Law: I think several people by the following names know
> something about the murder of my wife: Mr. Ward Pen-
> ning, Mr. Joe La Londe, Mr. Vilas Lant, Mr. John
> Deskins, Mr. John Vekich, Mr. Red Silen, Mr. I.E.
> Mosier.

> Manley: Anyone else?

> Law: I have been threatened by Mr. Jack Clark, whose editor-
> ial for the return of the good old days when they mur-
> dered and got away with it, as they are now...

> Manley: Do you suspect Mr. Clark of being implicated in this
> murder, do you?

Law: I said I think Mr. Clark possibly knows something about it. I said 'possibly.' I say it would be at least partially honest for your office to subpoena Mr. Joe Schneider, who has threatened me; I say that even perhaps Mr. Russell Mack, who has incited toward hatred for me for some years, who has said, 'I must go,' who has slandered and vilified me and my friends, the labor movement, democracy, civil liberties, who states that wrecking of the Finn Hall is understandable, when it should be known that the murder of my wife is a natural sequence of such condoned vigilantism that you and your office and the local police have overlooked.

Manley: Now, Mr. Law, we are giving you a good bit of scope here.

Law: You took a lot of scope. [88]

Law was interrogated by the coroner and the prosecutor for a day and a half, but the ten men that he had named were not called before the inquest recessed.

The Grays Harbor Civil Rights Committee and a National Grays Harbor Committee were formed to gather support for Dick Law. Among the sponsors of the national committee were Theodore Dreiser, Rockwell Kent, Carey McWilliams, Philip Murray, Michael J. Quill, George Seldes and Richard Wright. U.S. attorney general Robert H. Jackson and the FBI met with a three-man delegation from the Pacific Northwest but found no ground upon which to intervene.

On March 20 and May 1, Goodman and Caughlan submitted requests for a resumption of the inquest; Coroner Stevenson denied both requests. When the inquest finally resumed on June 3, the ten men named by Dick Law had been summonsed to appear. Seven of the ten appeared in court; one was in San Francisco, two were ill. By the time the third man was called to testify, both Goodman and Caughlan had been ejected from the courtroom for their attempts to cross-examine. When Law persisted in asking questions, he was routinely cut short by the coroner.

It is clear that a double standard was operating. Law had been interrogated by the prosecutor and coroner for a day and a half; the interrogation of all seven suspects (plus William Baker, who had been summonsed to testify for A.R. "Red" Silen who was in San Francisco) was accomplished in a few hours. That done, the inquest was closed.

The jurors deliberated for only a half-hour and concluded that "Mrs. Laura Law was murdered by an assailant or assailants unknown to us." The attempt to manipulate the inquest to frame Dick Law for murder had collapsed; to that extent, the verdict was a victory for the Left in Grays Harbor and the Pacific Northwest. But it was really a shallow victory for if, as the left averred, there was a direct tie between the Better Business Builders, the Finn Hall incident, and Laura Law's murder, the vigilantes had escaped unscathed.[89]

The coroner's inquest could have exposed that tie and uncovered the truth about vigilante activity in Grays Harbor, but too many questions about the ten men named by Law had been allowed to go unanswered. Two had been arrested for the Finn Hall wreckage, but found not guilty. One had a key to Law's house. Another had been charged by Law with theft of IWA union funds. Two constantly editorialized against Law in their newspapers. All disagreed with Law's union activities. Many had argued bitterly with him and several had threatened him.

The terror campaign that had been launched against the IWA had lasting effects. Laura Law's death was seen by her friends as a political assassination. The fear that one's family or friends might be next on the "hit list" made radical political involvements a high risk; the fear that one might be framed for the crime was even more disconcerting. The activities in Aberdeen, Washington were only the first sign that the IWA's future would not be decided at the ballot box. During the next year, the United States immigration service and National CIO would make the decisive interventions in the IWA's history.[90]

4
Harold Pritchett's Deportation and the Intervention of the CIO's National Office

The referendum election following the 1939 convention had returned the IWA's left-wing administration to office but the workers' votes were not the last word on the matter. The Columbia River District Council (CRDC) greeted the election results with accusations that the election had been improperly conducted even though its candidate, Al Hartung, and the regional CIO directors for Oregon and Washington had approved the procedures beforehand. Michael Widman from the national CIO office subsequently investigated the election and declared its results valid.[1]

The CRDC's objections might have been dismissed as sour grapes if it had not been for the Council's next move: it recommended to its locals that they cease paying the per capita tax for the International's organizing program. The Council was quite willing to starve the parent body rather than provide funds for the union to extend into Canada and the South where new locals could be influenced by the left-wing International office. The per capita recommendation was in fact an implicit ultimatum to the International, with the CRDC threatening to withdraw or alternately risking suspension.[2] However, Oregon State CIO director William Dalrymple, who usually supported the CRDC, advised the locals to pay the tax and to remain within the IWA. After the largest local, Portland Local 3, voted to table the District's recommendations on the tax, Hartung and Helmick withdrew their threat to take the CRDC out of the International, but the matter of paying the tax was not resolved for several more months.[3]

The United States Immigration Service Intervenes

Since his first entry into the United States in 1933, Harold Pritchett had been issued twenty-two ninety-day visitor's permits and nine extensions. Ten days before the founding meeting of the IWA in Tacoma, Washington in July 1937, he had been excluded "as an immigrant alien not in possession of an unexpired immigration visa, and as a person likely to become a public charge." At the time, Pritchett had been receiving $100 monthly from the British Columbia Council of Lumber and Sawmill Workers and he was president of the Federation of Woodworkers. Fortunately, Frances Perkins, President Roosevelt's Secretary of Labor, had intervened on Pritchett's behalf and he had been allowed to enter the United States temporarily as a visitor. Ten days before the First IWA Convention in December 1937, Pritchett had again been ordered to leave the country because his permit had expired. Again, he had been granted an extension. From 1937 on, the IWA president had resided more or less continuously in Seattle, only returning to British Columbia to visit his family or to avoid breaking the immigration laws when his permits expired.[4]

In February 1938, immigration authorities had again denied Pritchett's request for a permanent visa which would have enabled the Pritchett family to establish residence in the United States and acquire citizenship. The authorities cited secret evidence which they claimed established his affiliation with the Communist Party of Canada. Several months later administration supporters formed a "National Committee to Gain Entry for Pritchett." In hopes of pressuring the authorities to reverse their decision, they distributed 50,000 glossy, four-page pamphlets which asked, "Shall 100,000 Wood Workers Be Deprived of Their Chosen Leader?" In an attempt to rebut the immigration authorities' contention that he was a Communist, the pamphlet quoted Pritchett as saying:

> I am not a Communist. We have Communists within the ranks of our organization, and I have no objection to this whatsoever, providing they live up to the constitution of the union. The employers decide the personnel of our membership in as much as they employ Democrats, Republicans, Communists, Socialists, and persons with other lines of political thought . . . [5]

Harold Pritchett had, in fact, been a Communist since the early 1930s, but if he had admitted it, he would have been promptly barred from the United States. At the same time, he would have been accused of committing perjury for he had signed a sworn affadivit that he was not a Communist.

His situation was even more complicated by the separation of jurisdiction which existed between the Immigration and Naturalization Department, which was under the Labor Department, and the Consular Service which came under the Department of State. The Department of Labor was responsible for issuing Pritchett's temporary visitor's permits while he was waiting for the Department of State to issue a regular visa which would have allowed him to become a resident of the United States.[6] However, the State Department continued to maintain that it had "certain secret evidence allegedly showing [Pritchett's] connection with the Communist Party."[7] But at numerous hearings in Vancouver and Seattle Pritchett was never confronted with any Communist Party financial statements showing payment of dues or contributions or membership cards he had signed. Instead, the hearings discussed Communists whom Pritchett might have known, statements by informants, and whether or not he had read or distributed various Party pamphlets. The 1935 funeral in Surrey, B.C. was referred to repeatedly.[8]

The lumber operators, the AFL's representatives and the CRDC all stood to gain if Pritchett was denied entry by the immigration authorities. If he was out of the way, employers could deal with a weakened, intimidated, less militant IWA, and the anti-administration struggle by the AFL and CRDC would be resolved. Former opposition members have confirmed that it was just "understood" that the CRDC would not support Pritchett's attempt to obtain a permanent visa, and that when the decision was announced, "the White Bloc was tickled to death he couldn't get in."[9]

A letter from Washington Congressman Warren Magnuson to Paul Coughlan, Pritchett's attorney, outlined the role of lumber operators during the immigration hearings:

Investigation has revealed that much of the pressure and objection to Mr. Pritchett entering the US has not emanated, as is commonly supposed, from the Seattle AF of L but has come from the American Consultate in Vancouver. They have been, I feel, influenced in their objections by many lumber companies operating both in the States and in British Columbia, but with smaller units in B.C. Those operators feel, therefore, that Pritchett, kept in Canada, will be a lesser thorn in their sides than were he allowed to come here. I have talked at great length about this to Mr. Simmons, of the State Department, who is handling this matter, and he has verified this opinion.[10]

Wayne Morse, the Dean of the University of Oregon School of Law and a Pacific Coast labor arbitrator — he was later the Senator from Oregon — believed that this last request by Pritchett for a visitor's visa was "...fair and reasonable..."[11] Consequently, five days before the final decision, he attempted to intercede on Pritchett's behalf with a letter to Secretary of State Cordell Hull. The letter read in part:

...I have become more or less familiar with the problems of the International Woodworkers of America. At the present time that Union is in negotiations for a new contract with the Employers on the West Coast. I think it must be admitted by all concerned that this union is very much in need of leadership and counsel of Mr. Pritchett at this time. Personally, I am satisfied that it would be in the interests of sound industrial peace to make it possible for this international union to have the benefit of Mr. Pritchett's leadership in the United States as well as in Canada.[12]

Pritchett's final application for a six-month "non-immigrant visa as a temporary visitor" was refused by the American Consul General in Vancouver, B.C. on August 22, 1940. According to the Consul, Pritchett had "...failed to establish non-immigrant status..." and his admission "...into the United States would be contrary to the public safety, as set forth in the [Immigration] Act of October 16, 1918, as amended by the Act of June 5, 1920."[13]

There was no further recourse for Pritchett. The U.S. government had successfully barred the elected president of the IWA from presiding over his union's convention. In his letter of resignation, Pritchett blamed the powerful interests of the business community, plus regional CIO directors William Dalrymple and Richard Francis, and Adolph Germer, the newly appointed IWA director of organization, all of whom had purposely intervened to oust the administration.[14]

The capricious treatment of Pritchett's immigration requests did not stop in 1941. Although he was president of the British Columbia IWA District #1 and an executive board member of the Vancouver Labor Council, he was refused admittance into the United States and thus prohibited from attending International Union conventions until the IWA convened in Vancouver, B.C. in September, 1944. The following year Pritchett's successful appeal enabled him to enter the United States intermittently until August, 1947. Then, even though the Labor Progressive Party of Canada, to which Pritchett belonged, expressly forbade the use or advocacy of violence, he was again excluded from the United States because of his membership in an organization that allegedly advocated the overthrow of the U.S. Government by force or violence.[15] That same year the entire left wing of the Canadian delegation was denied permission to attend the IWA Convention. In all, throughout the 1940s, the United States Government regularly denied twenty-seven Canadian IWA members the right to enter the country on union business.

It appears clear that the government of the United States was determined to bar Pritchett from the country because of his effectiveness as a union organizer; his membership in the Communist Party was only an expedient technicality. Had Pritchett and his supporters foreseen the future better they may have approached his defence differently. Instead of arguing the legalities of his case, perhaps they could have mobilized rank-and-file actions in his support. At the time, however, there were more pressing matters (such as the fight with the AFL), so they turned to people such as Wayne Morse to influence the immigration authorities' decision.

Pritchett's forced resignation as International President had some compensations, however, for it allowed him to organize full-time in Canada, where it appeared possible to build a solidly left-wing IWA district that would be beyond the reach of the Dies Committee and the other reactionary influences of the U.S. In other words, it was at least theoretically possible to outflank the repressive apparatus of any one nation state by building a strong and balanced international union.

"Mr. CIO"

The loss of Harold Pritchett was a severe blow to the progressive forces in the IWA. He was a powerful speaker, an experienced and respected organizer and a hard worker. Harvey Levenstein, in his recent book, *Communism, Anti-Communism and the CIO*, adds that Pritchett was probably the IWA's "only leftist leader with considerable personal appeal."[16] But the Communists' problems in the IWA were deeper than the loss of an individual: nearly one hundred years of political and economic fragmentation was setting the limits on what the IWA leadership could accomplish. The industry's workers had been seriously divided since the 1935 strike, the Carpenters' campaign to defeat the IWA had developed into a crusade behind which the AFL was willing to throw much of its considerable weight, and the Columbia River District Council, still witholding its dues in protest of the 1939 elections, operated as a fifth column within the ranks of the lumber workers.

Opposed by the enormous resources of the AFL, the IWA had turned to its own parent body, the CIO, for help. This may have been a miscalculation on the part of the Pritchett leadership, but it isn't clear whether there were any other choices. It seems likely that the left simply expected John L. Lewis to take a principled position of support against the AFL and against the CRDC which had been a reluctant joiner in the industrial union movement from the beginning. But Lewis had other ideas. He had tolerated the presence of Communists in CIO affiliates because they were the best organizers he could find. But now that the unions were built, it was time to return to a more mainstream course. Perhaps recognizing the vulnerability of the IWA, Lewis chose this time to initiate "what was to become the first successful attempt to oust Communists from control of an important union."[17]

The IWA's history at this point is important to an overall understanding of the CIO-CCL period. It has been conventionally argued that the CIO National Office took a hands-off attitude toward Communists in its affiliates until the post-World War II years.[18] But recently, Harvey Levenstein has challenged that conclusion and our detailed examination of the IWA from 1940 to 1942 confirms his thesis.

In early 1940, John L. Lewis had sent Michael Widman, CIO Director of Organization, to the Northwest to survey the situation. When he returned to Washington, Widman had recommended that a National CIO organizing director be sent west immediately who would:

> have charge and direction of the organization staff [and] approve all bills in connection with the organization campaign before some are paid by the International Secretary. He shall hire as many organizers as the fund will permit, who shall be employed on a trial basis. He shall direct their transfer from place to place as conditions may warrant such transfers. He may discharge organizers for cause. The final decisions on organization campaign problems that may arise from time to time must rest in the hands of the director.[19]

Lewis appointed Adolph Germer to head the IWA's organizing program, ostensibly so that neutral organizers could be hired and their activities carried out without regard to the union's internal politics.

The role of Adolph Germer in the IWA, the CIO, and labor in general has not been given the attention it deserves. Germer's name turns up in the accounts of many major labor struggles of the twentieth century. He was present at the Ludlow massacre in 1914; he went to jail with Eugene Debs during World War I; he was national secretary of the Socialist Party between 1916 and 1918, and is credited with enlisting police help in keeping pro-communist delegates out of the Socialist Party's 1919 Chicago convention. Germer led a United Mine Workers faction in opposition to John L. Lewis during the 1920s and 1930s before joining with Lewis in the formation of the CIO; he showed up in the organizing drives of the Rubber Workers, the Auto Workers and the Oil Workers and, for more than ten years, was a central figure in the IWA. In his later years, Germer often took over the offices of CIO Vice-President and Director of Organization when Allan Haywood was absent. In 1946–47 he represented the CIO in Paris as Assistant Secretary-General of the World Federation of Trade Unions in charge of Colonial Affairs. Germer was, in fact, once introduced as "Mr. CIO."

By the late 1930s, Germer's reputation as the CIO's chief troubleshooter in areas of Communist activity had already been established. Len De Caux, one-time editor of the *CIO News*, remembered Germer as a "rigidly sectarian anti-leftist." Harold Pritchett remembered Germer as a "bitter old socialist who made a career out of going to jail with Eugene Debs." Germer, moreover, represented an emerging strata of labor leaders who saw rank-and-file militancy and ideological groupings as sources of instability that impaired the collective bargaining process. To "hold the rank and file in line," according to Lorin Lee Cary, "[Germer] labored incessantly to augment the authority of union leaders." Nowhere did Germer better fulfill his role as troubleshooter than in the IWA.[20] As a result, he became the pivotal character in the consolidation of opposition to the Communist faction within the IWA.

Within a short time after Germer arrived in Seattle in June 1940, he had become a close union confidant of Al Hartung, President of the Columbia River District Council. If Harold Pritchett, Karly Larsen and the Midwest Finns typified the immigrant working class element within the union movement, then Hartung represented the opposite — the agrarian, upwardly mobile native American. He was born on a Wisconsin farm and grew up with a dream of going west where the country was still expanding. By 1929, he had traveled between the Midwest and the West Coast several times; he had been a professional boxer and wrestling promoter and tried farming with a sister in Montana. He had tended bar and worked in the woods. In 1929 he took a job as a foreman in a Sedro Woolley, Washington, logging camp and for the next four years worked as a foreman in camps in Washington and Oregon. And when radicals began organizing the unemployed and the Trade Union Unity League, Hartung had been busy earning a reputation as a "slave-driving s.o.b." as a logging supervisor for the Clark and Wilson Company near Vernonia, Oregon.[21]

Future IWA executives Worth Lowery and James Fadling, (who would become the first two IWA International presidents elected by the anticommunist White Bloc, while Hartung would be the third) and Carl Winn, (elected vice president during the White Bloc sweep in 1941) had interests and backgrounds similar to Hartung's, and they also soon aligned with Germer. Lowery and Fadling had come to the Northwest from Oklahoma; Winn, like Hartung, was from the Midwest. Winn's father had been an Idaho "camp push," the lumber industry's equivalent to the "overseer" on a slave plantation. Hartung, Lowery and Winn had been (or claim to have been) Wobblies; Winn's brother had been a leader of the anti-communist faction of the IWW. The Germer-Hartung inner circle was thus well suited to do battle with Harold Pritchett and the radical leadership of the IWA.[22]

On June 19, 1940, Pritchett had communicated to IWA district councils and locals that Germer was on duty. According to Pritchett's memo the organizing campaign would begin as the Widman report to Lewis had stated, "*with the exception* that the National Representative of the CIO office will direct the organizational drive *in cooperation with* the four International officers, and *counter-sign all checks* on the separate organizational fund with our International secretary, B.J. McCarty"[23] (emphases added). The underscored portions of Pritchett's memo, when compared to Widman's recommendations to Lewis, indicate that the IWA leaders had a quite different concept of the CIO's involvement in their affairs than did Widman and Lewis. Essentially, it was the IWA's autonomy in organizing activities that was at stake, but within a few months it would be apparent that Lewis had gotten a foot in the door and was not going to remove it.

The first step in the campaign was a referendum which called for a fifty-cent assessment to support the organizing drive planned for the period August 1940 through January 1941. The anti-administration, anti-Pritchett forces in the IWA had blocked previous assessments for fear they would only beget a stronger administration, but it was Germer's hope that with the CIO in charge of the organizing, the anti-administration forces would support the referendum. But before the referendum was even completed, a fight began over who would be hired as organizers.

Control of the 1940 organizing program was perceived by both factions to be critical, for the uneven political development within the industry's workforce meant that *where* organizing activities were conducted would also determine *who*, in political terms, would be brought into the union first. This building process would be the key to eventual control and direction of the International. Political affiliations soon became the principal criteria by which organizers were selected. When John Gibson, secretary-treasurer of the Michigan State Industrial Union Council wrote to Germer recommending Matt Savola for an organizing position in the Michigan area, Germer replied:

> With reference to Matt Savola...you understand the situation in the wood-workers union. It is divided into two warring groups. In other words, we have the same difficulties in the Woodworkers that we had, and now still have, in the auto workers — CP vs. Anti-CP. Out here there is a feeling that Matt is definitely and actively lined up with the CPs.
>
> ...you said something about [Savola] going with the Communist Party as an organizer. This information and the expressed feeling out here among the Anti-CPs prompts me to write on this subject.
>
> I wish you would give your confidential opinion.[24]

By August 9 Germer had received lists of recommended organizers from both factions.[25] In consultation with Oregon CIO director Dalrymple and "one of the other boys in the Columbia River District," Germer ascertained that all but two of the names submitted by Pritchett were "labeled" — a term indicating some association with the Communist Party.[26]

On August 16, Germer informed Orton and McCarty that he had no intention of submitting a list of proposed organizers to Harold Pritchett for his approval, and that he had decided to interview five men from a list submitted by the Columbia River District Council.

The left was well aware of Germer's potential to load the appointments of new organizers in favor of the opposition. None of the men recommended by Karly Larsen, president of District Council Two in northern Washington — William Wallace, James Freeman, Robert Bernethy, Robert Blakely, Ralph Nelson and Harold Hanson — were to be interviewed by Germer. Larsen's recommendations only provoked Germer to ask: "Are these recommendations yours [meaning from Larsen and the left] or were they produced by convention action?"[27] The Midwest District threatened to shut off their per capita assessment to the International if Matt Savola was not hired as an organizer. Pritchett, too, objected: three of the proposed organizers "[were] persons with no former organizational experience or record of achievement and lean towards comfort to the opposition against the program and policy of the IWA and the CIO."[28]

Germer interpreted the left's responses to his plans as evidence that "the fifth column [the communists are] splendidly at work. . . I was not far wrong when I predicted that there would be plenty of sabotaging during this campaign." On August 22, 1940, the very day Pritchett's final visa request was denied by the immigration authorities, Germer put eight new organizers — four of them men rejected by the International — on the IWA payroll.[29] With Pritchett banned, thanks to the Immigration Service, and new organizers who favored the opposition hired, Germer was positioned to reshape the rest of the IWA. Just three months after his arrival in the Pacific Northwest, he was in a position to ply his skills at two important gatherings — the Washington State Industrial Union Council Convention and the Fourth Annual Convention of the IWA. Germer entered this new arena with a revealing and important vote of confidence from his superiors in Washington, D.C. On September 9, 1940, Michael Widman assured him:

> I have noted carefully our reports for the past few weeks [and] am wholeheartedly in accord with your actions to the smallest detail. . . I can only suggest that you keep battling away at them [the Communists]. Your judgement seems sound and I know your experience will guide you in the best interests of all concerned.[30]

The Industrial Union Council Convention

The IUC was a state-wide umbrella organization designed to coordinate the activities of all CIO unions in Washington State. From its inception, the IUC's executive board was alleged to be dominated by pro-communist labor leaders from the Seattle area. National CIO leaders therefore viewed the IUC as another "front organization for the Communist Party" and launched an attack on the Council when it convened for its third annual meeting on September 20, 1940.

Prior to the convention, Germer and Richard Francis, regional CIO director for Washington State, had attended IWA local meetings to encourage the rank and file to oust the elected IUC leaders at the coming convention. Francis also acquired all the IUC's per capita records and submitted a list of dues-paying locals and members to the Convention's credentials committee as a pressure tactic to get favorable delegates accredited.

When Francis and Germer arrived at the convention, however, they could see that delegates were not being accepted in accordance with the criteria that Francis had established. Members of the IWA and the Fishermen's and Cannery Workers' unions were entitled to dues exoneration because of their frequently seasonal work. Germer and Francis had hoped the full-dues and per-capita regulations would exclude delegates from both these unions because they were politically too radical for the regional directors' liking. The left-wing leaders in charge of the convention naturally wanted these same loggers and fishermen admitted as they counted on their militant support. The leadership also wanted to reassert the principle of industrial unionism that *all* workers — even those who were temporarily unemployed — had a right to representation in their respective union organizations.[31]

The fight over the seating of the delegates raged within the credentials committee and on the convention floor for two days. On the third day, following acceptance of the credentials committee report by the convention delegates, Germer addressed the convention and explained that "the convention could not conduct business in a legal manner as the [convention] call [had been] illegally issued." A motion was then made to adjourn the convention and have John L. Lewis "take full charge of the Washington State Industrial Union Council. . . until another convention can be properly called."[32]

A flurry of letters, phone calls and telegrams between John L. Lewis and his subordinates in Washington State concluded on September 26 when the CIO president appointed a "committee to assume charge of the Washington State Industrial Union Council and to administer its functions."[33] That committee consisted of Richard Francis, William Dalrymple, Adolph Germer, Harry Tucker from the Aberdeen IWA, Jack Bell from the Mine, Mill and Smelter Workers local in Tacoma and John Bator from Roslyn, Washington. It would have been difficult to select a more conservative, pro-opposition committee to guide the Industrial Union Council.

Germer had moved quickly. The opposition not only seized control of the Washington State Industrial Union Council, but Harry Tucker, a tried and true anti-communist from Aberdeen, was selected secretary-treasurer of the committee which would oversee the IUC and administer its affairs. While the Washington State Industrial Union Council was not the IWA, the national CIO and Adolph Germer had, nevertheless, gained an important springboard from which to continue their crusade against the left-wing unionists in the Pacific Northwest.[34]

The Fourth Annual IWA Convention

The momentum gained by Adolph Germer at the Washington State Industrial Union Convention carried over into the IWA's convention at Aberdeen, October 7–12, 1940. With Harold Pritchett barred from the United States, O.M. "Mickey" Orton, IWA First Vice-President and a Pritchett supporter, assumed the convention chair. Germer directed the anti-communist delegation from his prominent seat in the visitors' gallery. On numerous occasions, White Bloc opposition delegates left the floor to seek advice from him, then returned with some procedural harassment for Orton.

The first feud of the convention erupted when Worth Lowery, IWA second vice-president and a leader of the opposition, asserted that the *Timberworker* had continued policies that were not in the best interests of the lumber workers. According to Lowery, some editorials created ". . . dissension and dissatisfaction" within the IWA by criticizing Don Helmick and supporting "causes and issues divorced from our problems as workers in the lumber industry."[35]

Lowery also accused the Communist Party of "meddling" in CIO affairs. As evidence, he cited *Sound Timber*, a leaflet issued in Portland by the Communist Party during August 1939, which stated that Party members would continue to support the lumber workers' struggle for better wages, hours, and working conditions.[36] Lowery concluded his address by recommending that the convention unanimously adopt a resolution demanding the Communist Party cease ". . . interfering in any way with the affairs of the IWA."[37]

Don Helmick, Al Hartung and Worth Lowery, supported by Adolph

Germer and the Oregon and Washington CIO directors, William Dalrymple and Richard Francis, were now prepared to oust the Communist leadership of the lumber workers union. Ralph Peoples, secretary of the Oregon State Industrial Union Council, introduced the resolution to condemn communist activities in the IWA.

> Resolved, the fourth Convention of the IWA go on record condemning these actions of the Communist Party and demanding in no uncertain terms that the Communist Party cease at once its interference in the affairs of the IWA.[38]

The next two days were devoted to the discussion of the resolution. Don Helmick expressed the opposition's feelings when he said:

> I am not going to dilly-dally about the expression of my opinion on the question of communism or any other of that rotten cancerous type.
>
> Don't duck it; don't crawl on your bellies. You can do all the conniving, the manipulating you want to do, but if you are an American and you believe in democracy, say so now. We know who these people [Communists] are... You've got them here.[39]

The resolution was overwhelmingly adopted, 251 to 5.

Still not satisfied, the CRDC delegates, for the third consecutive convention, proposed a constitutional amendment that would exclude all Communists and Fascists from the IWA. The anti-communist faction cited the precedent of such a clause in the United Mine Workers' constitution, but William Riley, an IWA executive board member from Alabama, pointed out that the UMW had never submitted that provision to its membership for ratification. Riley also called attention to other undemocratic practices of the UMW such as the appointment of district presidents by its international president.[40]

Immediately, William Dalrymple, who was seated in the visitors gallery, charged to the microphone and demanded the right to answer Riley's criticism of the Mine Workers Union. Orton, who was chairing the convention, gaveled Dalrymple down. "It is not within the authority of any regional director to interrupt the proceedings," of an IWA convention, he told the convention. Denied the floor, Dalrymple stalked out of the convention accompanied by a handful of supporters.[41]

The anti-communist constitutional amendment was again defeated, 134 to 124, with the Columbia River District Council casting 54 of the dissenting votes.

Immediately after the convention closed, Orton telegraphed John L. Lewis to protest Dalrymple's behavior at the convention. Three days later Lewis wired his response: "... Regret to learn of controversy within your organization. I have no information on which to base any conclusion respecting actions of any CIO representative."[42] If the administration had narrowly carried the convention, the opposition clearly carried John L. Lewis.

The CRDC called a "rump" session after the regular convention adjourned, with Al Hartung presiding over 100 delegates who claimed to represent seventy-five percent of the dues-paying membership. They petitioned John L. Lewis to investigate the "internal strife" at the convention allegedly created by the administration officers. A resolution adopted by the "rump" session accused the administration of attempting to purge lumber workers who opposed communism, of discriminating against lumber workers who failed to vote with the administration, and of promoting the interests of the administration in the *Timberworker*.[43]

To counter the proceedings of the "rump" convention, the IWA executive board met on October 17, five days after the regular convention adjourned. The meeting was primarily concerned with the activities of Adolph Germer. Since his arrival in the Pacific Northwest, he had successfully organized lumber workers around Longview and Aberdeen, but he had also committed himself to the opposition and undermined the administration's control over many IWA locals.[44] Furthermore, he had encouraged William Dalrymple and Richard Francis to oppose the IWA administration at the recent convention. The IWA executive board decided to send Mickey Orton to Washington, D.C. in an attempt to persuade John L. Lewis to remove Germer. The board wanted Germer replaced by a union official who was "... non-partisan and less interfering in the administration and internal affairs [of the IWA] and a more capable organizer." Lewis disregarded the request, and a few days later Germer was told by Lewis' number-one assistant, Allen Haywood, to "rest assured insofar as the CIO National Office is concerned, you are the Director of Organizaton, IWA."[45]

Longview and Aberdeen

The war between Germer and the IWA continued unabated into October 1940, when the Southern Washington District Council became the battleground. Two issues converged to fuel the fight. One was Germer's appointment of George Brown and Chet Dusten as organizers for Longview Local 36. The local had protested the appointments when they were made in August and had elected a five-man committee to escort the two out of town, but Brown and Dusten had stayed and followed Germer's plan to organize the more conservative sawmill workers rather than the loggers in order to shift the political balance of Local 36. The two men were also accused of paying workers to join the local in order to oust the left-wing leadership. The presence of Brown and Dusten was a sore that festered for weeks. Then in September when Germer assigned National CIO representative Gola Whitlow to the local, the situation was further aggravated. The members saw Whitlow as Germer's stooge, and it was only a matter of time before the Longview Local exploded.[46]

On December 20, the day logging operations were shut down for Christmas, the executive board of Local 36 convened an evening meeting. Whitlow and Brown appeared outside the meeting hall with several supporters and demanded admittance. When they were refused, they remained outside shouting threats at the executive board members and pounding on the door. When the board members attempted to leave the building, they were challenged by the gang outside.

"I can lick any one of you son-of-bitches," boasted Freddy Thomason, one of the disrupters.

Thomason then threw a punch at Bob Williams, the local's business agent, and this touched off a general melee in which at least one person was injured.[47] The tension and bitterness left by the fight virtually immobilized the local during the holidays, so that no attempt was made to resolve the dispute until the first meeting of the new year.

On January 10, 1941, after Local President Kenneth Deckert had adjourned the local's meeting, one of Germer's supporters, Charles Kumler, picked up the gavel and reconvened the meeting. A motion was passed abolishing the office of business agent, which was held by left-winger Bob Williams. The legality of Kumler's action was taken to court, where it was upheld. The court further declared that Williams and Deckert had been holding Local 36 property illegally, and then granted an injunction restraining the International from activity in the local.

In the midst of this struggle in Longview, dissension arose over Southern Washington's contract negotiations. The lumber industry, spurred by production demands for the coming war, had begun to pull out of its depression. In turn, the IWA and AFL increased their demands for higher wages. In February 1940, the IWA International policy committee had made plans to meet with employers for the first industry-wide negotiations; they had selected a committee composed of two members each from the five districts involved, plus the CIO regional director and the International officers.[48]

In May, however, these industry-wide talks had been scuttled when the individual employers had insisted upon their rights to accept or reject the contract negotiated by their representative, the Lumberman's Industrial

Relations Committee. Separate agreements had been subsequently concluded on September 10 by the Columbia River District Council, and on September 18 by the "Twin Districts" of Northern Washington and Grays-Willapa Harbors. But two of the Twin District locals, Raymond and Aberdeen, had refused to ratify the contract and instead signed separate contracts with the employers.[49]

Matters were further complicated by the AFL's negotiations with plywood operators, which had been taking place at the same time. When these negotiations failed, the AFL struck mills in Tacoma and Snoqualmie and then attempted to spread the strike to mills in Everett where the IWA had agreements. At this point, the leaders of the IWA's Columbia River District Council had realized that the Everett situation provided the perfect opportunity for the launching of their own attack on the left-led Northern Washington District Council, and they agreed to cooperate with the AFL in a fight against the Northern Washington District. Apparently the CRDC hoped that an AFL victory in the Everett area would curtail further growth of the Northern Washington District, thus altering the political balance between Northern Washington and the CRDC to the latter's advantage.

On November 9 the CRDC and the AFL exchanged representatives to attend each other's meetings. They met again on December 1 in Centralia to "extend their collaboration."[50]

Disgusted with these shenanigans, the loggers in the Aberdeen/Raymond Local decided to pull out of the IWA. They had always been a source of union militancy and progressive political ideas, and with an independent local they hoped to remain so. They applied for a separate charter in order to insulate their local from Germer's influence. Unfortunately, this step split the ranks of the leftists, as some of them did not agree it should have been taken. It also isolated them from other woodworkers in the Aberdeen area, leaving Germer's influence in those locals uncontested. Following the loggers' departure, Local 30 joined Local 36 in Longview in the White Bloc camp, thus completing Germer's take-over of the Southern Washington District.[51]

The open collaboration of the IWA's White Bloc with the AFL made it clear that the struggles within the IWA entailed more than abstract ideological differences and personal power plays. The policies and politics of the two factions were contradictory. The International's Communist leaders sought to build a union along industrial principles which would be a force for social change within the lumber industry, while the Columbia River District Council and Germer were willing to subordinate those ends to the restoration of "respectable" business unionism akin to that of the AFL. The fate of the IWA as an industrial union was thus at stake when it squared off with Germer and the White Bloc for the show-down fight.

Germer was not acting alone, however; he had received mandates from his superiors on September 9 and October 21. And when Philip Murray had become National CIO President in November, Germer's campaign against the Communists had gained even greater backing. In December, Germer's office had received "instructions from Murray [which] indicate clearly that he plans a cleaning-up on the Reds in no unmistakable manner." Germer had thus been given a "green light" to go after the IWA's leaders.[52]

Fighting for its life, the IWA executive board voted 11–4 on December 5, 1940 to expel Adolph Germer and ordered him to vacate his office immediately.[53] The board characterized Germer's policies as "disastrous and disruptive" and criticized other regional CIO officers for supporting the anti-administration movement.[54] The controversy between the IWA and the CIO's representatives was summarized in the letter notifying Germer of his expulsion:

> ...Richard Francis of Washington and William Dalrymple of Oregon,...in collaboration with Adolph Germer, have pursued the policy of injecting themselves into the internal affairs of the IWA.

> In selecting men as organizers, those who were supposedly aligned with certain individuals within the IWA were given preference. Upon hearsay, rumors and whispered slander, men who have contributed much to the IWA were ruled out as organizers by Germer...The counsel and advice of elected officers of IWA went unsolicited by Germer. The motive of Germer was obvious. It was to make more secure, at the expense of the IWA membership and its elected officers, his own precarious place in the labor movement and, in cooperation with Dalrymple and Francis, to further personal ends.[55]

With that, virtually all organizing work came to a standstill. Both sides began documenting their positions on the events that had taken place during the fall of 1940.

On Trial: Adolph Germer or the IWA?

On January 21, 1941, CIO President Philip Murray informed the IWA that an investigation of all the charges leveled against Germer would be conducted by a committee composed of Reid Robinson, president of the International Union of Mine, Mill and Smelter Workers, J.C. Lewis, who was a special CIO representative, and the United Rubber Workers' president, Sherman Dalrymple. (Lewis, from the United Mine Workers, had been appointed administrator of the Washington State Industrial Union Council in a move by John L. Lewis to take that council out of the hands of left-wing leaders.)

With Dalrymple as the presiding officer, the committee began its work on January 18 with a formal hearing in Seattle. Prior to considering the actual charges against Germer, the committee heard arguments on the scope of the committee's mandate. The committee's duty was clearly to investigate the charge against Germer, which obviously made him the defendant, but Germer argued that the committee should expand its scope to investigate the entire controversy between the pro-communist International leaders of the IWA and the minority anti-communist faction based in the Columbia River District Council. Germer's definition of the committee's mission would have allowed him to put communism and the IWA's leaders on trial and, although the committee would not agree to that agenda, he had taken the first step in manipulating the proceedings in that direction.[56]

The charges against Germer fell into three categories: political discrimination in the hiring and assignment of organizers, an active campaign to discredit the International officers, and the initiation of overt disruption in the political affairs of IWA locals. International President O.M. Orton was the first witness against Germer. He laid out a general indictment against Germer for having aligned himself with those IWA members who were originally opposed to the CIO and who were presently opposed to the organizing program, of making "accusations and vilifying statements" against the IWA's officers which included allegations of misappropriation of funds, and of alleging a conspiracy between Orton, Harold Pritchett and Harry Bridges to control the West Coast CIO through a merger of the IWA and the ILWU.[57]

Following Orton, witness after witness testified to Germer's disruption of and interference in the business of the union's international and local affairs. Howard Dyer, president of the Klamath District, testified that Germer had accused Pritchett and Longshore leader Harry Bridges of being communists; Edson Stalcup from Local 3-30 in Lebam, Washington reported that "Germer appeared at [his] local union meeting...to start a campaign of red-baiting [against] the International officers" and advised members to "clean house in the state CIO organization"; Dick Law told the committee about Germer's appearance before the Columbia River District Council where Germer had told the members, "this [is]...a fight between this district and...the Communist Party."[58]

The International's case against Germer for discrimination in hiring organizers focused first on the hiring of Tommy Beaird as an organizer in the Alabama District, where the IWA southern drive was being spearheaded. William Riley, an International Executive Board member from Alabama, presented the International's case against Germer's hiring of Beaird. William Mitch, Alabama State CIO President and Yelverton Cowherd, regional CIO director in Alabama had originally recommended that Riley be hired as an organizer, but Germer had refused on the pretext that Riley held a paid position on the Executive Board of the IWA.[59] With Riley's approval, Mitch and Cowherd then recommended that Oscar Pruit be hired. The recommendation was made, however, shortly before the 1940 International convention to which both Riley and Beaird went as delegates. At the convention Beaird won Germer's favor with a good display of red-baiting, while Riley earned his dislike for criticizing the United Mine Workers Union for having a clause in its constitution forbidding Communists from holding union office, and for being generally undemocratic in its internal operation. Germer was fiercely defensive about the UMW, his "home" union, but he was even more angry because he was trying to use the experience of mine workers to support his contention that the IWA needed a communist exclusion clause in its constitution as well. Shortly after the convention, Germer hired Beaird despite the objections of Riley.[60]

Germer had hired four organizers in the Northwest over the objections of the IWA leaders, and by the time of the trial two of them, Joe Clark and Walter Funk, had been accused of causing serious disruptions. Clark, an outspoken critic of the International officers, had tried to set up "Committees for Getting the Local Union on the Right Trail" and as a result had been ousted from the Klamath District.

Funk, who had been assigned to the Eugene, Oregon area was accused of a number of disruptive acts including the beating of William Harris, a member of the Coos Bay District and a supporter of the International officers. Funk also accompanied Germer's other appointees to District conventions and executive board meetings when Germer requested a "stacked" meeting. Funk and others from the Eugene area had even traveled to Seattle to attend an Executive Board meeting while on the payroll as organizers. By the end of 1940, Germer's appointed organizers were considered little more than a roving goon squad which he could summon on a moment's notice.[61] Germer's encouragement of strongarm tactics in the Longview area were also laid before the committee, and the International tied the events in Longview to Germer's larger mission of getting rid of the leftist element in the West Coast CIO.

The largest CIO unions on the West Coast were the IWA and the Longshoremen's union; both were considered left-wing by national CIO leaders. Soon after his arrival, Germer had linked the relationship between the two unions to the future direction of the entire CIO, and expressed concern over the fact that the ILWU was going to assess its membership fifteen cents a month to assist a joint IWA/ILWU organizing program. He also noted there were hints of an amalgamation of the *Timberworker* and the *Voice of the Federation** into a single Pacific Coast paper. At stake was control of the entire West Coast CIO movement, and Germer realized that the communists would have gained strength if the two unions merged.[62]

Harry Bridges appears to have been the major target in the CIO's campaign against such a merger, but the level of his organization in California made it more feasible for the CIO to move through the back door, using the anti-communist climate in Portland, Oregon, as its wedge. The Portland-based Columbia River District Council therefore became a staging area from which the CIO national office gained access to the CIO State Industrial Union Councils in Oregon and Washington. Germer's post as Director of Organizing for the IWA allowed him to coordinate the campaign.[63] Meanwhile, the suspicions of Germer and the National CIO appeared confirmed when Bridges spoke to the IWA's convention in October of 1940, suggesting that the executive boards of the IWA and ILWU map out a program for organizing woodworkers. "This," commented Germer, "was interpreted by many to mean that [Bridges] plans to take over the International Woodworkers of America."[64]

Following that convention, a series of meetings between IWA President Mickey Orton and Longshore leaders had been held in Longview. At a meeting on October 25, Orton reportedly suggested that "the Longshoremen be given jurisdiction over the maintenance in the mills. . . because of the internal friction in the IWA"[65] Had that step actually been taken it would possibly have been the first of many leading to the eventual merger of the IWA and ILWU into a coast-wide industrial union under left-wing leadership, but Germer's organizers had pulled the pin on this collaboration effort when they broke up the Longview Local's Executive Board meeting on December 20, 1940, and seized control of the Local.

Germer's Case

On the sixth day of the trial, February 3, 1941, Adolph Germer began the presentation of his own case. The resemblance between his trial strategy and that of the government in its proceedings against alien radicalism was striking, and while the likeness may have been coincidental, the fact that in the Fall of 1940 after the International began moves to oust him, Germer

---

* The *Voice of the Federation* was the paper of the Maritime Federation of the Pacific, an organization comprised of eleven unions including the ILWU, Inland Boatman's Union, and Marine Cooks & Stewards.

had requested copies of the Dies House Un-American Activities Committee hearings suggests that his course may have been more consciously charted.[66]

Germer led off with three "character witnesses": his assistant and fellow CIO representative Harvey Fremming, Richard Francis, who was the CIO Director for the State of Washington, and Worth Lowery, an International IWA vice-president and leader of the White Bloc. Then, borrowing a page from the Dies Committee proceedings, Germer relied upon three ex-Communist Party members to establish the "fact" that several IWA leaders were members of, or sympathetic to, the Communist Party. One of the witnesses was Nat Honig who had recently been fired as the editor of the Party newspaper, *The Western Workers.*[67] Although what being a Communist meant in terms of trade union performance was less important than the association itself (in the strategy followed by Germer), the specifics of "communist" activity cited by the witnesses are worth noting. D.L. Shamely, vice-president of Local 3-30 and a star witness for Germer, testified that communists had made local union meetings the scenes of intense political debate between supporters of the Washington Commonwealth Federation (a farm-labor alliance) and the Republican Party. Shamely complained that local meetings sometimes lasted until 1:00 a.m., and cited the introduction of education programs in shop committee meetings as a "communist" offense.[68] Shamely's attempt to discredit as "communist" what others defended as free speech, open debate, political and democratic unionism, and an informed rank and file, indicated how far he, Germer, and their supporters were from the principles of industrial unionism as understood and upheld by the Communists. That Germer's defense would appeal to conservative union instincts in this way indicated his transformation (in Lorin Lee Cary's words) from labor agitator to labor professional.[69] As the trial progressed, it became apparent that Germer would be able to amass enough evidence of Communist Party involvement in the IWA to justify everyting he had done. After that, as Executive Board member Dick Law said, "the hearing became an inquest, and accuser the accused."[70]

James Fadling, who would later become International President, made it explicit that the White Bloc did not want the committee to investigate Germer but, rather, the Communist Party.[71] But he inadvertently helped the case against Germer by pointing out that Germer had attacked the Communist Party before the Party attacked him. "When Brother Germer come [sic] out here, before he made his stand against Communism, he was a fine man, an able leader," said Fadling, but "as quick as he became outspoken on the question of Communism" he came under attack by the

Communists.[72] Another of Germer's witnesses, Harry Tucker, testified that, "No attack was made, to my knowledge, on Brother Germer until he unloaded on the Communist Party. . ." Here was an admission that Germer had brought an anti-communist campaign to the IWA and had precipitated the confrontation between the Left and Right. Yet Fadling, Tucker, Germer and the White Bloc followers, apparently blinded by their own anti-communism, could not see that in the context of the proceedings, their statements constituted evidence of Germer's guilt — not his innocence.

When committee member Reid Robinson questioned whether the committee was investigating the role of the Communist Party in the IWA or the activities of Adolph Germer, he was answered by Germer himself.[73]

> . . . suppose it is brought out here that the same persons who attended a certain fraction meeting three years ago, four years ago, two years ago or one year ago are the same persons, and where the intimate affairs of the [IWA] were discussed in these fraction meetings, are the same people, some of them who have been conducting this campaign of slander against me, I think we have a right to show the tie-up.[74]

Robinson also asked that testimony alleging that Harold Pritchett and O.M. Orton were members of the Communist Party be stricken from the record because the witness, Harry Tucker, had offered no evidence to prove it. Robinson argued that the allegations were "deliberately done. . . to prejudice the case in favor of Adolph Germer, and [have] no relevancy. . . in this case."[75]

After Robinson's objections were overruled, Germer's tactics began to pay big dividends. From mid-afternoon until 11:15 p.m. on February 4, 1941, the committee was caught up in a debate over who was or was not a Communist, and whether this was or was not a relevant question. For hours the committee lost sight of the charges against Germer and turned the proceedings into a witchhunt.

I Confess

The trial of Adolph Germer came to an end on February 6. Germer himself took the stand as the last witness. He reviewed his eight-month involvement in the IWA's organizing program and the controversies surrounding his policies, all the while lauding the gains made by the people he had hired. Throughout his lengthy presentation Germer called attention to the presence of Communists in the IWA and highlighted his campaign against them. Then, he climaxed his testimony by reading from *I Confess* by Benjamin Gitlow. Gitlow, a former Communist Party National Secretary and two-time running mate of William Z. Foster on the Communist Party presidential ticket, had left the Party in a faction fight during the late 1920s. As did many ex-Communists, he had turned rabidly anti-communist and wrote venomous denunciations of his former comrades. Germer read passages about alleged Communist Party conspiracies to disrupt the LaFollette presidential campaign in 1924, and their conspiracy to drive John L. Lewis out of the labor movement and seize power in the United Mine Workers. As Germer continued, the story of "Moscow gold" being used to finance the U.S. Communist Party was accepted as a truism.

Dramatically, Germer linked up Gitlow's portrayal of Communist conspiracies to events in the IWA. "[The present] campaign of slander and falsehoods against all who are opposed to the Communist Party's interference in the affairs of the CIO generally and the IWA in particular," said Germer, "is perfectly in line with the methods employed by the Communist Party in earlier years, so clearly described by Ben Gitlow."[76]

Germer continued:

> . . . I cannot escape the conclusion that behind the curtain there is the directing hand of the Communist Party as Benjamin Gitlow so clearly pointed out in his *I Confess*. I have known them ever since the Russian Revolution of October 1917 and before, and although their pretentions vary somewhat from time to time, their objectives are the same in every detail.

But if Germer's remarks were anticipatory of the Cold War that was still six years in the future, the response they drew from Ilmar Koivunen were even more prophetic. Koivunen, the Finnish logger from the midwest and an IWA International vice-president, had been the chief spokesman for the International during the trial. He saw not only what Germer was about, but with remarkable insight cut through Germer's sanctimonious veil and exposed his red-baiting tactics:

> . . . I just want to say that to an unbiased jury that Germer's testimony in itself should convict any man on the basis of the charges that we have made. It is a very clear indication that Germer has become a great believer in *I Confess* and now is starting to carry out his ideas and intents, instead of coming to the IWA as an unbiased, unprejudiced person to. . . organize the woodworkers into the IWA Germer has launched himself in a crusade according to the rules set down by *I Confess*, and I think that anyone reading over. . . the final remarks of Adolph Germer in his testimony, should bring clearly to your minds that he did not have the intent in his mind, when he first set out. . . to organize.

> . . . Germer has come out in this territory in the hopes of helping the man who wrote *I Confess* to carry out his crusade [by linking] up everybody and anybody indiscriminately of being communists or fellow travelers. . .[77]

No cross-examination of Germer was made by the IWA, leaving Koivunen's words to punctuate the proceedings. Koivunen was of course correct in his assertion that an unbiased reading of the record would confirm Germer's guilt, but the U.S. labor movement was embarking on an era which would vindicate Germer, not Koivunen.

The exonerated Germer was left in charge of the IWA's organizing program until Februay 1944, when he was promoted to CIO Director of the West. In 1946 he was chosen by Philip Murray as a man "who would restrain the communists. . . ." and appointed to the post of CIO ambassador to the World Federation of Trade Unions.

Koivunen, on the other hand, left the IWA a few years later to become a longshoreman in Coos Bay, Oregon. He retired a working man, and his efforts to halt the tide of hysteria that had rolled over North America remained unrecognized for almost forty years.[78]

The Oregon State CIO

The factionalism within the IWA directly affected the proceedings of the third Oregon State CIO Convention held in Eugene from January 31 to February 3, 1941. David Fowler, regional CIO Director from Oklahoma and a close friend of Germer and John L. Lewis, addressed the convention, urging the Oregon CIO to adopt the position of the United Mine Workers and ban all members who joined the Communist Party.[79]

William Dalrymple, Oregon CIO director, got up to defend the Columbia River District Council's actions against the left-wing leadership of the IWA, but in the year since the 1940 convention, relations between Dalrymple and the Portland Industrial Union Council (PIUC) had deteriorated. Dalrymple regarded left-wing lumber workers and PIUC spokesmen as liabilities to the Oregon CIO movement, and he questioned the propriety of sending a PIUC delegate to the Emergency Peace Mobilization meeting in Chicago. While Dalrymple supported President Roosevelt, the PIUC followed John L. Lewis and declined to give Roosevelt unqualified endorsement for a third term. Furthermore, the PIUC's newspaper, the *Labor New Dealer*, which had been the official organ of the state CIO since 1937, had rebuked Dalrymple for exceeding his authority in supporting the ouster of the IWA's left wing.[80]

On October 28, 1940, the PIUC had sent a letter to John L. Lewis informing him that the "uncooperative and dictatorial attitude of Dalrymple might destroy the Portland CIO movement by decreasing rank-and-file support."[81] The letter contained testimony which Francis Murnane, delegate from the Plywood and Veneer Workers Union, had presented to the PIUC, summarizing the split between the Council and Dalrymple:

Brother Murnane charged that Brother Dalrymple had disgraced the CIO by his red-baiting and alien-baiting tirade... Brother Murnane further charged and he was in a position to know, that Brothers Francis, Dalyrmple and Germer had meddled in the internal affairs of his International union [the IWA] and that they had directed the strategy of the well-known "opposition" bloc that had done so much to hinder the IWA and CIO in general. [82]

At the convention, the Oregon CIO movement suffered a severe setback when the PIUC's publication, *Labor New Dealer*, was repudiated by the convention. The paper was accused of carrying out "smear campaigns" against the IWA's White Bloc and promoting "disorganization instead of organization" within the Oregon CIO movement. The Council adopted the *CIO News,* the national CIO newspaper, as its official organ. [83]

While the State CIO Convention failed to adopt the anti-communist resolution suggested by Fowler, the radical forces within the industrial labor movement had been successfully challenged. With the *Labor New Dealer* silenced, the influence of the left-wing Portland Industrial Union Council was greatly diminished. William Dalrymple was now in effective control of the Oregon CIO movement.

The First Split: the CIO Woodworkers Organizing Committee

The trial committee had left Germer in charge of the IWA's organizing program. With the exception of Worth Lowery, all the International's officers and many of the locals protested that the trial had been a kangaroo court, and they repudiated its decision to retain Germer. However, on February 20, 1941 Lowery formed a "CIO Organizing Committee", based in the Columbia River District Council and funded by the CIO. It published its own newspaper, *The Woodworker,* and embarked on its own organizing program under the direction of Germer. This organizing committee was designed to be a political weapon for use against the elected left-wing IWA leaders. [84]

The White Bloc campaign was endorsed by the national CIO on March 15, 1941, when Germer was informed of:

... a conference with Phil Murray... [where we] discussed... the organizational drive, that is to carry it on in the favorable places to our cause and by this method take [care] of the IWA problem in their next convention, as there isn't any way to take care of them under the CIO constitution, only in convention to kick them out.

... with this Spokane and Willamette Valley organized, we should be able to take the COMRATS in the next IWA convention in great style... (emphasis in the original). [85]

The White Bloc organizers now set out to fill the IWA's ranks with members of their own political stripe and they found them in the Willamette Valley. Running south from Portland for a hundred miles to the Eugene/Springfield vicinity, the valley had long been a stronghold of conservatism. It was in the lower reaches of this valley that the first Oregon Trail pioneers had settled small plots of land. The small sawmill lumber industry which had matured there prior to the IWW era contrasted sharply with the heavily capitalized industry around Puget Sound which had received the socialist Scandinavian workers at the turn of the century. The conservative political character of the Willamette Valley work force had thus been established early, providing a suitable opening for the AFL, and until the IWA appeared in 1941, for the Industrial Employees Union, the successor to the government-created Loyal Legion of Loggers and Lumbermen (4-L).[86]

The White Bloc's organizing victories in the Willamette Valley further polarized the union. By April 1, 1941, Germer reported that the "breach between the two groups was beyond repair." Locals which were aligned with the CIO Woodworkers Organizing Committee withdrew their support from the IWA and were suspended; the CIO, however, stood behind them and threatened to set up separate district councils to receive the suspended locals and to issue them direct industrial union charters. The California Provisional Council was added as part of this campaign.[87]

In late July, Allan Haywood, CIO Director of Organization, attempted to established a Unity Committee between the two factions. President Orton, International Secretary Bertal McCarty, Columbia River District Council President Al Hartung, and Executive Board Member Ed McSorely from the Columbia River District Council were called to Washington to forge a peace pact.

Haywood proposed constitutional changes that would give the White Bloc more power. Since 1937 locals had been represented at International conventions on the basis of two delegates for the first 100 members in the local, and one for each additional 300 or major fraction thereof up to 1,000.[88] Roll-call votes had been conducted on the basis of one vote per delegate regardless of the size of the local the delegate represented. This structure also prevailed within the locals led by the left wing.[89]

Haywood's proposal called for one delegate for each local up to 100 members, and one for each additional 300 members. The roll-call procedure would be changed to allow each delegate to vote a proportionate share of the membership of the local union he or she represented.

The changes had very clear political implications. The one-unit, one-vote rule on roll-call votes which the Communists favored increased the relative amount of power for the smallest, poorest and most recently organized locals. These were the locals most likely to consist of loggers and unskilled workers who had traditionally been discriminated against by the craft dominated AFL unions, and who had been the most radical historically. They were, therefore, the units whose need was the greatest, but whose political inclinations were the most progressive.

This voting structure, however, ran counter to the popular business expression that "those who pay the freight, drive the train." The White Bloc controlled the three largest locals in the International and the largest district council. By voting its membership strength on roll-call votes, the White Bloc directly controlled almost one-fourth of the total votes. Haywood's method of allocating delegates would give the White Bloc three additional delegates while the British Columbia District, which was emerging as a left-wing power, would lose three delegates. Three Washington State left-wing strongholds, Bellingham, Sultan, and Everett, would lose delegates. Every local in the International, in fact, would lose a delegate except for four — three of which were staunchly anti-communist.

Besides roll-call provisions for convention votes, representation on the International Executive Board came up for change. The Four-Man Committee proposed that each of the four officers of the International be given a vote on roll calls equal to the total strength of the International divided by the number of districts, twelve. At the time, the Columbia River District Council controlled twenty-five percent of the votes. Since they intended to exclude communists from membership the new provisions for roll-call votes would give the anti-communist faction an additional vote on the executive council equal to four-twelfths, or one-third, of the total strength of the International.[90]

Exactly how the agreement to change the constitution in this manner was arrived at is unclear. Walter Galenson says the left "conceded" the changes at the Unity Committee's Washington meeting, and he reasons that this was in line with the Communist Party's adoption of popular front politics after Germany's June 1941 invasion of the Soviet Union.[91] This is an important point because studies of many similar instances of Party-line changes have supported the conclusion that the Communist Party weakened itself by placing international priorities ahead of domestic concerns. In the IWA case, however, the popular front explanation is not totally satisfactory. There was no apparent souring of rank-and-file sentiment toward the Communist Party because of its line changes; in fact, of the five locals transferring allegiance during the year, three went from the White Bloc to the Red Bloc.

The only available record of the Unity Committee's meeting with Haywood is Germer's diary. In it, he recorded on July 30, the last day of the meeting, that Orton would not agree to the changes. And again, on August 7, Germer noted that there was no agreement.[92]

As convention time approached, Germer's correspondence reveals that he expected the issue to be resolved at the convention, and he began campaigning among friendly delegates, leaning particularly hard on the Alabama District Council. On the first roll call vote of the convention sixteen of seventeen Alabama delegates voted with the White Bloc.[93] After the first day of the 1941 convention, Germer wrote in his diary:

> The Whites were clearly in control from the start. . . . Haywood saved the day for them by his membership roll call proposal.[94]

It may have been the case that communist popular front policies softened opposition to Haywood's proposal, but in Germer's mind the Party's position was a less important factor than the CIO's manipulation of the balance of political forces within the IWA.

When the 1941 convention met, forty-seven new locals had been added since the previous year and twenty-eight of them voted with the Columbia River District Council on the first roll-call vote. No new locals had been added in British Columbia. The votes added to the White Bloc by the organizing program, the changes in the voting procedure, the deportation of Harold Pritchett and the ideological effect of the CIO's anti-left crusade gave the anti-communists the clout to pass a resolution banning Communist Party members from the union. The amendment (see Appendix I) was ratified by the membership after the convention, and it remained in effect until 1973.[95]

In an attempt to unify the opposing forces, Orton decided not to run for re-election as IWA President. Harold Evans, business agent for the Olympia plywood locals, replaced Orton as the left-wing candidate; Orton ran for vice-president. Worth Lowery, the incumbent second vice-president and White Bloc leader, opposed Evans for the presidency. Left wing Secretary-Treasurer Bertal McCarty was opposed by Ed Benedict.

The referendum election following the convention elected an entirely new slate of officers to the International: Worth Lowery, president; James Fadling and Carl Winn, vice-presidents; and Ed Benedict, Secretary-treasurer. Al Hartung, Columbia River District Council President and two-time loser to Harold Pritchett for the International presidency, was added to the CIO payroll as Germer's assistant. The union's new leaders reaffirmed their ties with the CIO and retained Adolph Germer to direct the IWA's organizing campaign.

Whether the indigenous conservatism represented by Hartung and his followers could have sustained itself in the absence of Adolph Germer and the CIO will never be known. The Trotskyist newspaper the *Militant* touted the "anti-Stalinism" of the Columbia River District Council as a progressive force in the IWA; but there is little evidence that the kind of sophisticated anti-communism organized by Trotskyists and social democrats in other CIO unions and in the IWA's British Columbia district would have developed in the Pacific Northwest. And it is clear from Germer's reputation as a factionalist in the Socialist Party and his ties to political movements in the East, that he was the real connection to those sectarian tendencies. It is also clear that he provided the technical guidance for the fight against the left and that he was responsible for the material resources that flowed from the national CIO.[96]

Although he had spent his life in socialist and trade union circles, he was cynical about working class politics. He and the White Bloc leaders were basically willing to accept the arrangement of class relations under capitalism and seek their own advancement within that framework. For them, a form of unionism that would be stable and assure them a comfortable role as spokesmen for workers was preferable to one which was part of a working class social movement that could transform society in a socialist direction. Germer thus became the conduit for the influence of an emerging strata of professional trade unionists within the CIO. Individuals like Hartung, Lowery and Winn were astute enough to parlay the combination into their own rise to power.

The left wing of the IWA, however, was by no means dead, and by the end of World War II, its success in organizing in British Columbia would bring it back as a power to be reckoned with at the International level.

5

The Left Rebuilds in British Columbia

By the 1940s, the conditions for organizing in British Columbia had become more favorable than in the U.S. Pacific Northwest, because the B.C. industry was more capital intensive and its workforce more exploited. In addition, the large B.C. mills concentrated their workers in mill towns and logging camps on islands and coastal inlets where the realities of their class existence were hard to escape.[1]

However, in 1940 the British Columbia District Council Number One of the IWA was in desperate straits. It had just suffered an eleven-month quarry workers' strike at Blubber Bay which had left the union with only a few hundred members and little money. It was therefore decided that the program to rebuild the district would focus first on the Lake Cowichan area of Vancouver Island and on the Queen Charlotte Islands.

There are 500 miles of ocean between Vancouver and the Queen Charlotte Islands "dotted with logging camps all up and down the coast." This stretch of water "was one of the toughest to navigate in the world, filled with islands, jutting rocks, racing tide-waters, rock mountains rising steeply from the sea, with no vestige of shelter for miles for a small boat in a storm."[2] To organize loggers in such rugged natural conditions, the IWA acquired the first boat in its colorful "Logger's Navy," now a legend in British Columbia.

The turning point in the early war years was the organization of the Chemainus mill, a holding of the Victoria Lumber and Manufacture Company.[3] Beginning in mid-summer 1941, IWA organizers Hjalmer Bergren and George Grafton leafleted and talked to mill workers, and in the spring of

103

1942, the union applied for and received certification under the wartime labor act. Within weeks, mills at Hillcrest and Youbou followed suit; Fraser Mills in New Westminster and the big mills in Vancouver were organized soon afterwards.[4]

In 1942, Harold Pritchett was elected president of the British Columbia District and helped lead the drive in the Queen Charlottes. The Islands were covered with Sitka spruce, a wood urgently needed in the war effort for the manufacture of airplanes, but although the British Columbia lumber operators were making millions on wartime production, "they refused to bargain for fair wages and conditions for the men who did the killing work of production."[5]

In the summer of 1942, the Queen Charlotte loggers told the union that they refused to continue working because conditions had become so bad; as a result, the Allison Company lost over half its crew of fallers and buckers.[6] Pritchett first considered breaking the union's wartime no-strike pledge, but instead, charged the companies with sabotaging war production. However, when the British Columbia Loggers Association, which spoke for the operators, refused to accept a ruling of the Federal Arbitration Board that it recognize the union (Local 1-71), 900 IWA members struck on October 2, 1943. The fourteen-day strike ended with "the unconditional surrender of the four major companies involved." The first IWA agreement with the British Columbia Loggers Association followed, as did recognition for the IWA in all camps and mills where it had won elections. Prior to this time, winning an election had not necessarily meant being recognized by the company, since Canada had no legislation comparable to the Americans' Wagner Act. With victories at Lake Cowichan and the Queen Charlottes, the IWA had changed the course of the British Columbia labor movement; now this left-wing power base would have to be reckoned with. But as the organizers left the islands for the mainland, the anti-communists prepared to challenge the left for control of the District.

The "Old Timers" at New Westminster

The first target of the mainland organizing drive was Fraser Mills, owned by the Canadian Western Lumber Company. Ten years after the mill's operators had confronted the Workers Unity League, Maillardville was still a company town, and the owners were still exploiting the Chinese, Japanese and East Indian immigrants who were housed in its company-built shacks.

The Fraser Mills local began as part of Vancouver Local 1-217, then during the Fraser Mills organizing drive in the fall of 1942, a group loyal to the union's District leadership and the Vancouver local formed Local 1-357 and elected Harold Pritchett, then secretary of the Vancouver Labour Council, as its president. At about the same time, a second leadership faction sprang up when some of the Fraser Mills workers struck for a wage increase. George Mitchell and Stewart Alsbury, both employees at Fraser Mills, tried to keep the men at work; then failing in that attempt, the two men – without union authorization – represented the strikers in negotiations with the company and the Labour Board.[7]

Mitchell and Alsbury were associated with a faction calling itself the "Old Timers Group" which issued a series of leaflets attacking the IWA B.C. District leaders. It accused Harold Pritchett of "using the IWA as a racket for his own personal profit" and called the IWA a "scheme to extend the influence of the Communist Party." Despite the deplorable conditions at Fraser Mills, the "Old Timers" leaflets contended the company was committed to the protection of "worthy employees." And, while Fraser Mills would not object to "a well governed union of its employees," no company could "reasonably be expected to willingly embrace the kind of an organization the IWA and its leadership has proved to be." The "Old Timers" encouraged the company to make no deal with the IWA.[8]

The membership of the "Old Timers Group" was never made completely known, but the similarity between its literature and that of the company was remarkable.[9] The District officers claimed the "Old Timers" group was only pushing the company line and trying to break the IWA organizing drive because of their political differences with Pritchett and the Communist Party. At the time, the Cooperative Commonwealth Federation (CCF), the forerunner of the present social democratic New Democratic Party (NDP), was locked in a bitter struggle with the Communist Party for leadership in the Province. Alsbury (whose brother Thomas was a leader of the CCF) and Mitchell were known to be CCF supporters. Alsbury had emerged as an opponent of left-wing union leadership as early as the 1931 strike at Fraser Mills,[10] and he eventually admitted being part of a dissident faction which had published a leaflet called "Union Facts" criticizing the B.C. District leadership.[11]

The anti-communist dissidents at Fraser Mills also had links to the IWA White Bloc faction across the border. Money for the "Old Timers" leaflets had come from the U.S. side.[12] International Secretary-Treasurer Ed Benedict was said to have gone "into Fraser Mills and talked to those who are and have been at all times in opposition to the leadership of [the] district." Claude Ballard, a White Bloc stalwart from the Portland area and a former International President, was reported to have been meeting with the B.C. dissidents, and the International's director of organization, George

Brown, was charged with having been "hoodwinked" into complicity with the dissidents.[13]

Thus, within months after the IWA had moved their organizing drive to the mainland, the effort to build a militant industrial union among British Columbia's woodworkers was bogged down in a fight with a fifth column opposed to the Communist leadership of the District. But the involvement of the International office and the anti-communist dissidents from the U.S. side of the border already foretold that the B.C. woodworkers would not be allowed to resolve the dispute on their own. Within months, the controversy at Fraser Mills would become a mere tributary in a flow of national and international political movements.

A Second Front: 'Trade-union Illiterates'

The British Columbia working class, its labor unions, and political formations had matured rapidly during the depression years. With maturation had come increasing sensitivity to the nuances of political strategy, tactics, goals and disagreements. But although socialism had become a goal common to many B.C. workers, a shared understanding of what socialism meant was less common. Differences on this issue irritated older wounds, for some Socialist Party stalwarts had been harboring grudges since the days when the Russian Revolution had drawn thousands of their comrades to the Soviet cause. The time for settlement of the old scores was nigh.

But if the social conditions at Fraser Mills provided the fertile ground for an anti-communist movement, the Cooperative Commonwealth Federation (CCF) provided the seeds, the nourishment and the caretakers. They made political hay out of the CIO's arrival in Canada. "Many CCFers, especially members of the Cooperative Commonwealth Youth Movement (CCYM), became CIO organizers. CCF units assisted striking workers by providing pickets and meeting places."[14] The CIO was called a magnificent opportunity, one the CCF "must not mess up."[15] Pat Conroy, Secretary-Treasurer of the Canadian Congress of Labour during the 1940s "regarded the young CCFers who had helped to organize the new CIO unions as 'trade-union illiterates'. . . whose primary motivation was not to build a labor movement but to capture it for the CCF."[16] Even so, Conroy and the other anti-communists in the Canadian industrial labor movement "valued the CCF's strength in the CCL as a bulwark against the Communist 'menace'."[17]

On the national scene, the CCF was encouraging political and labor organizations to become "affiliated" with the party in order to allow them to participate on an official basis in CCF and electoral affairs. The IWA's 1943 convention voted to affiliate with the CCF, but the terms of union affiliation, adopted by the party's Provincial Executive on January 28, the same day the CCF received the IWA's request, required that "every delegate from [an] affiliated trade union...must not be a member or active supporter of any political party or political organization other than the CCF."[18] When the IWA's District leadership rejected these terms of affiliation, the CCF leaders within the union, including John Ulinder of Ladysmith who was president of the Cowichan-Newcastle CCF district, and Lloyd Whalen who was chairman of the CCF Trade Union Committee, began to attack the IWA leaders.[19] This move was just part of a drive launched by the national CCF and the Steelworkers' Union to "rid British Columbia labor of Communist domination."[20] The CCF also turned down bids for affiliation from the Mine, Mill and Smelter Workers, and from the Shipyard Workers — the two other largest unions in the province. And when the Labor Progressive Party (LPP) in British Columbia applied for affiliation on September 4, 1943, it was denied because of the LPP's adherence to the principles of democratic centralism and its supposed subservience to Soviet foreign policy.[21]

On October 18, national CCF leader David Lewis advised the party's Trade Union Committee in British Columbia to

> concentrate its efforts on wresting as many of the locals as possible from Communist control...Shaky Robertson [Steel official, is being sent to B.C.]...with instructions...to start the ball rolling.[22]

Robertson, who was joined by Eileen Tallman, also from the Steelworkers Union, had orders to make the anti-communist drive a "joint Steel-CCF enterprise." He was described as an ex-communist who "was so violent and single-minded in his anti-Communism that even [Steelworkers' President Charles] Millard was horrified."[23] With Robertson's arrival, the history of the CCF struggle against the Communist Party converged with the continuing crusade of the IWA's anti-communist International officers against the B.C. District leaders.

The Malaspina Conspiracy and the Trial of John Ulinder

On November 2, 1944 a letter circulated by John Ulinder, a member of IWA Local 1-80 in Duncan, announced plans to "dislodge the LPP domination of the IWA." "A committee has been set up," he wrote, "to clean house in the IWA." The letter promised expense money for "supporters" who attended a special meeting to be held at the Malaspina Hotel in Nanaimo in ten days' time.[24] When George Grafton, business agent of Local 1-80, heard about the meeting he went to Nanaimo with local members Owen Brown and Fred Wilson, but Ulinder refused to allow them into the meeting. With Ulinder was Shaky Robertson.[25]

Grafton filed charges against Ulinder for disruption and for "knowingly promoting a secret and unauthorized meeting" at which "plans were laid to unseat officers and expel members of the union by improper means." Ulinder was further charged with "planning with other members of the Union and with persons not members of the Union to cause officers of the Union to lose their positions without just cause."[26]

Ulinder was tried by Local 1-80 on January 14, 1945 and found guilty.[27] At the trial Fred Olkovich, Local 1-363, testified that he had attended the meeting called by Ulinder. The purpose of the meeting, said Olkovich, was

> to set up committees in different camps, locals and sub-locals, to spread propaganda and unseat the present officers on the Executive, to bring out a strike policy, agitate the members in disrupting the present Executive, to bring out parts of the present contracts and the work that the Executive has done and to emphasize the weak points, to undermine the present Executive, to set up sub-locals, to break up the Local System, make it easier for the organization to take hold, to disrupt union meetings by walk-outs until such time as the membership of this organization was strong enough to take over.[28]

At the meeting Olkovich was also informed that "the International officers would supply certain information if it was deemed necessary to unseat President Pritchett and would also supply information that would start this organization (i.e. the group meeting at the Malaspina Hotel) in a fight."[29] It was clear, Olkovich said, "that those present had organized themselves into a group which had links with the CCF Party."[30] The strategy, he said, was "...to break up the larger locals of the IWA [B.C. District] into smaller units, each of which would be chartered as a local by the International." Additionally, it was planned "to attend union meetings and there cause disruption and antagonism" and "to antagonize the Union's present leadership by questioning it on the no-strike policy."[31] Finally, the Ulinder-Robertson meeting had discussed "the possibility of having all IWA members who were under the influence of [the White Bloc] leave the Union and resign their membership with a view to weakening the Union and confusing and disrupting it."[32]

Ulinder's conviction by his local was sustained by the District Council on March 7, 1945, but overturned by the International Executive two months later,[33] after he contended that the local had convicted him on insufficient evidence and that he had not received an impartial appeals hearing before the District Council. He went beyond his own case in his appeal, leveling charges against the B.C. District leadership for "collaboration with our employers," "defeatism," and "betrayal of trust" in negotiating the 1945 contract. He charged his own local officers with a host of violations of democratic principles including "toy[ing] with our funds."[34] International Board member William Harris labeled the condemnation of the District leadership "irrelevant," but on a 9 to 5 roll-call vote the Board voted to "sustain the appeal of John Ulinder and reverse the decision of the union local and District Council."[35]

While Ulinder's case was being settled, the program of disruption planned by the nine-person Malaspina meeting was being carried out. A series of leaflets in the name of the IWA "rank and file" were issued. The leaflets, entitled "Union Facts" and "The Undercut," attacked the IWA District leadership as "puppets, whose strings are pulled by a distant boss." Invoking the authority of the "rank and file" numerous times, the leaflets exhorted IWA members to regain union control from the "fakers," "stooges," and "sellouts," who were in control of the District.[36]

Attempts were made to break up the large IWA locals. Petitions were received from the Chemainus and Youbou sub-locals which asked for separation from Local 1-80 because the Duncan headquarters of the local was being "used to further the political ambitions of a minority group." The Chemainus petition had seven signatures, including that of one Clarence Sharp, allegedly a member of Pratt's detective agency, a leading union-busting organization in British Columbia.[37]

The petitions were considered by the International Executive Board at its August 5, 1945 meeting. Director of Organization George Brown cited the reasons for the petitions as the long distance involved for the sub-locals' participation in local politics, their dissatisfaction with the District's newspaper, the *B.C. Lumber Worker*, the no-strike policy and the general political leadership of District. "The District Council leadership up there has been furthering one political party," said Brown. "It's pretty hard to convince the members of the sub-locals that they're non-partisan when their own local union, that is the headquarters at Duncan, carried signs in front of the local union office 'Vote LPP' and the secretary of 1-80, Will Killeen, acted as agent for the candidate running for office under the banner of the Labor Progressive Party."[38]

But board member Nigel Morgan questioned the validity of the names on the petitions and defended the sub-local structure.

> The main reason that we formed a local and adopted a sub-local set-up is for the same reason of my own local. My own local stretches about a thousand miles along the coast line; it takes in all the B.C. Upper Coast. The reason it does so is because there are no roads. The only way you can get to the camp is by boat,

and they all ship out of Vancouver. The only mail contract is through town. It's a matter of administration. You have those all broken up into separate little sub-locals so that they can conduct their business. The cost would be so great and the problems of covering them — well, you have to maintain boats and you have to maintain cars.[39]

The International ended this incident by denying the Youbou and Chemainus petitions, but it called for an investigation of Local 1-80's internal activities.[40]

Organizers Fired

The third aspect of the Malaspina-CCF strategy called for the placing of its own people in union organizing positions so that newly organized locals would get off on the right foot — that is, the CCF's anti-communist political foot. The IWA's B.C. District, which was controlled by the LPP, offered few possibilities for the hiring of CCF organizers. Besides, the most experienced and proven organizers — men like Ernie Dalskog, Mike Freylinger and Tom MacDonald — were LPP supporters and available for further assignments. The CCF, therefore, turned for help in gaining access to organizing positions to the IWA's International office.

The hiring and firing of organizers soon became the lightning-rod issue of the fight. The first fired was Jack Greenall, secretary of the District Council.[41] Greenall had been hired as an organizer by Adolph Germer and had continued on when George Brown became Germer's replacement. Brown, however, charged that Greenall had interfered in a Local 1-357 election and fired him. The specifics of the charge were that when two men running for local office vilified the District Council leadership, the local began to question the two candidates' backgrounds. Greenall told a local meeting that "one [candidate] had been a special policeman during the [1932] long-shoremen's strike and that the other one had gone through the picket line during that strike."[42] The real issue, of course, was whether or not Brown, acting on behalf of the anti-communist International leaders, was attempting to aid the opposition to the District's left-wing leadership.

The battle for control of the organizing program continued when a drive was opened in the British Columbia interior in August of 1944. The first local, 1-405, was established in September,[43] then early in 1945, District 1 requested that George Brown transfer three organizers — Hjalmar Bergren, Tom MacDonald and Mike Freylinger — to the interior to assist in the drive. Bergren, who had led earlier campaigns on Vancouver Island, was considered "the most competent and popular organizer in B.C." MacDonald also had impressive credentials, having organized the Chemainus Mill; he had also been instrumental in organizing the Industrial Timber Mills, as well as mills in Victoria, Alberni and Vancouver. Freylinger had "carried the banner of the IWA through the difficult and trying times in the Queen Charlotte Islands" and had been "chosen to organize and consolidate the workers in the Fraser Valley."[44]

Brown, however, under the terms of the IWA's organizing agreement with the CIO, was not obligated to accept the recommendations and instead appointed Mike Sekora, Ralph New and Nick Kaptey. All three were relative newcomers to the union and were known to be opponents of the District's Communist leadership. Sekora and New had been employed at the B.C. Fir and Cedar Company and had only joined the IWA in August and July respectively. Sekora had been a delegate to a recent CCF convention; he had written a letter to the International's newspaper accusing District 1 President Harold Pritchett and International Executive Board member Nigel Morgan of not supporting the political program of the Canadian Congress of Labour (and the CCF). According to Sekora they had sold "the workers to the capitalists and the bosses."

The B.C. District office "opposed the appointments on the grounds that these men had demonstrated no organizational ability," that their appointment "would only lead to division and disruption among the members they were intending to organize and further, could only be expected to sow disruption within the IWA in B.C."[45] In addition, it was reported that Brown's organizers "didn't intend to concentrate on logging camps but rather were going to organize the mills first" and thereby establish small locals which would be financially dependent on the whims of the International organizers.[46] The District proposed a compromise slate of organizers comprised of Al Parkin, Mel Fulton and Bergren — the former two because they were from the communities being organized and Bergren because he was the District's outstanding organizer. Brown rejected the compromise.

On April 25 the B.C. District Council convened a special executive meeting to discuss the question of organizers in the interior. They voted to send one of their own officers to the interior to "[map] out an organizational program" and to protest Brown's appointments at the next International Executive Board meeting.[47] Nigel Morgan, International Executive Board member from District 1, carried the District's protest to the International and asked for a policy of "cooperation between the District and International instead of disunity and disruption." He expressed concern that Bergren, Freylinger and MacDonald would be laid off because the International's organizing funds could not support them plus the three men Brown had just hired.[48] The Board also heard a protest from a New Westminster local over the placement of an International organizer, George Mitchell. Local 1-363 contended that its operation was already well organized and it had not been consulted on the appointment.[49]

Brown defended his appointments to the interior, saying he would not "send any of the organizers that had been working in B.C. into the interior for the simple reason that the labor movement in Canada had developed into a political organization rather than a trade union movement." He thought it was "better to send new people in there rather than to carry the fight into that part of the country."[50] Brown further contended that Bergren, Freylinger and MacDonald were still on the payroll and would remain so.[51] The Board voted approval of Brown's decision and ordered Morgan to convey the board's policy to Local 1-405.

Following the International Board's decision, the B.C. District held a special delegate meeting on May 20 in Nanaimo. At the meeting, with International President Claude Ballard present, Pritchett pleaded for unity.

> We want to cooperate; we want to cooperate with the International officers... and nothing else. We are not concerned whether you eat spinach or cauliflower, whether you drink beer or straight water; whether you belong to the CFF, the LPP, the Liberals or the Technocrats. We are not concerned whether you are Jew or Gentile, East Indian or just plain Canadian. We have a job to do that requires the greatest degree of unity and anybody who gets in the way of that unity is going to get their toes stepped on...[52]

The delegates were angry, however, and voted to condemn and vigorously protest

> the...appointment of these International lackeys and demand their immediate removal and substitution of men who have proven their ability in organizing this District and in building up our union into a powerful and respected institution.[53]

International President Ballard proposed that the District and the International go to the CIO for resolution of the conflict.[54] This peace meeting was scuttled, however, when U.S. Immigration authorities refused entrance to the Canadian IWA leaders.[55]

The culmination of the fight over the organizing program came when Brown, despite his promise to retain the three left-wingers, fired Freylinger in mid-July for writing a letter inviting another union member to a meeting of the Labor Progressive Party. This letter was allegedly passed to George Brown by T.C. McKenzie, a vice-president of the local, who subsequently moved on to full-time work for the CCF. Brown appointed Mike Sekora to replace Freylinger.[56]

The events of the early 1940s are critical to this examination of the IWA in three ways. First, it can be seen from these events that the source of anti-communism among B.C. woodworkers cannot be so easily attributed to provincial nativism as suggested by author Iving Abella. In fact, if one wanted to vulgarize the record, it would be much easier to establish that Communist unionism was the predominant "native" tendency in the mid-1940s and that anti-communism was brought to B.C. by outsiders — CCF party leaders and right-wing American IWA leaders. But this formulation would still be an over-simplification; the uneven development of B.C. industry, the heterogeneous nature of its immigrant population, and the disparity in social conditions between work in the woods and work in urban mills had generated a variety of political attitudes. In short, both communism and anti-communism were rooted in B.C. class relations.

Second, the evidence confirms that the issues around which the struggle against the Left turned were not indigenous to lumber workers in British Columbia; the breakup of large locals and the agitation around the wartime no-strike pledge were issues introduced to the rank-and-file through leaflets circulated by CCF organizers and paid for with money raised outside the province. The controversy which surrounded the no-strike pledge was unusually spurious, because while the District's leaders promoted productivity plans in support of the war against fascism, they also supported strike activity when it was necessary. For example, the *B.C. Lumber Worker* had called for support of striking miners in December 1941 and for striking steelworkers in January 1943. Most importantly, the District was solidly behind the October 1943 strike by the Queen Charlotte Islands loggers which resulted in union recognition and a master contract. There is no evidence that the rank and file was actually opposed to the concept of the no-strike pledge.

Third, by resorting to undemocratic union methods and red-baiting, the dissidents were in fact admitting that rank-and-file support for the Communist leadership rendered ineffective any attempt to remove the radical IWA leadership by democratic processes. In other words, as of 1945, the very actions of the anti-communists constitute evidence that the Communists did not isolate themselves from the rank and file when they supported a wartime no-strike pledge and that they were not guilty of "undemocratic leadership" as some historians have alleged.

The 1946 Strike: "25-40 Union Security."

Early in 1946 the Canadian government announced its intention to maintain wartime wage controls and fixed five cents per hour as the maximum allowable wage increase. The employers, represented by R.V. Stuart Research Limited, opened negotiations on March 21 with a five-cent offer. When the union refused it, the offer was raised to twelve-and-a-half cents on condition that the IWA drop its demands for the forty-hour week, union security and the dues checkoff.[57]

Through the month of April the two sides remained deadlocked. Then on May 7, the IWA District Executive Board issued its first province-wide strike call for 11:00 am May 15, and raised the slogan "25-40 Union Security" — a twenty-five cent per hour wage increase, the forty-hour week, dues checkoff and other union security provisions. There was no strike-breaking because there were no scabs — the strike was that solid. Mass parades of pickets involved the public in the union struggle. When the Vancouver City Council rejected the union's request to solicit funds on a "tag day," strikers defied the authorities and collected $4,000 from Saturday shoppers.

The thirty-seven-day strike reached its peak when 3,000 strikers marched on the provincial capital in Victoria; within a week the government intervened to bring the strike to an end. Dominion Labor Minister Humphrey Mitchell informed the IWA that Gordon Sell, Esq., would henceforth control those plants and logging camps involved in the manufacturing of wooden containers: "Order requires that operators will open their mills at 12 noon, June 19 and that employees shall return to work. Rates of wages will be the same as those which were in effect when employees stopped work."[58] Chief Justice Gordon Sloan was appointed to arbitrate a wage agreement. The government's pretext for this partial seizure of the industry was the onset of British Columbia's interior fruit harvest. Without adequate supplies of wooden crates the harvest would rot in the fields and the strike's effect would ripple still farther across Canada.

With some workers forced back to work and others still out on strike, the IWA moved quickly to settle. On June 19 the executive board decided by unanimous roll-call vote to recommend a return to work as soon as possible. News of the settlement "was received by many rank-and-file members with some misgivings."[59] Union security was only partially obtained by the removal of the wartime no-strike requirement, and the wage hike fell ten cents short of the twenty-five cents demanded. Loggers, however, gained a forty-hour week for the second half of the contract, and an industry-wide contract covered the interior as well as coastal operations.

Bigger gains had been made on the organizing front, however. Ten thousand workers had joined the IWA B.C. District during the strike for a total of 27,000 members. With that growth the IWA became not only the largest union in B.C., but one of the largest in Canada.[60]

The Resurgence of the Left

As the strike had progressed, it had temporarily deflected the energies of the anti-communist crusaders and strengthened the left-wing presence in the union. This factor plus the strength of its organizing success during the strike helped the left wing to capture three International offices in the 1946 referendum elections. Karly Larsen from Washington, Ed Laux from Oregon and Jack Greenall from British Columbia defeated right-wing incumbents for vice-president, secretary-treasurer and trustee positions, respectively. British Columbia provided the bulk of the left-wing votes while the Columbia River District Council again accounted for the lion's share of the anti-communist vote (see Table VI).

The election was significant in two ways. First, it refuted the contention that the Communist Party's support for the war against fascism and for the wartime no-strike pledge in industry had turned the rank and file against union leaders associated with the Party. In fact, the documents and leaflets submitted as evidence in the trial of John Ulinder reveal that the no-strike pledge and the alleged Moscow connections of the Communist Party were issues within the union only because the Trotskyists and CCF used them in their attempt to bait and discredit the Communists. The 1946 elections proved that the leaflets issued in the name of the "rank and file" had been frauds, and that the rank and file did not support the activities of a handful of anti-communist dissidents.

Second, the vote was quite literally a red flag to the White Bloc, touching off a renewed attack on left-wing leaders in British Columbia and northern Washington. Adolph Germer, by this time a national CIO representative, wrote that a "certain group [was] again coming to the front and trying to take over." He accused his followers in the IWA of acting like "children" and warned them that they "will give them [the radicals] the organization."[61]

At this point, it appeared that the IWA's left-wing unionism was a Humpty Dumpty in reverse: all the King's horses and all the King's men couldn't take it apart. Despite the efforts of the CIO's organizers in the Pacific Northwest and the CLC's organizers in British Columbia, despite the red-baiting campaign of the CCF and the disorganizing done in British Columbian locals by the International, and despite the abrogation of democracy at all levels, the rank and file's support for its radical leaders was unshaken.

But if the King's horses and men couldn't get the job done, the King would. Soon after the 1946 IWA elections, the U.S. Government stepped in, just as it had in 1940 when it deported Harold Pritchett.

TABLE VI   **Results of Membership Vote for IWA International Officers, 1946**

	1st Vice President		Secretary-Treasurer		Trustee	
	Ballard	Larsen	Benedict	Laux	Mitchell	Greenall
District 1 (British Columbia)	737	5,205	933	4,950	599	5,277
District 2 (NW Washington)	574	3,543	848	3,239	878	3,184
District 3 (SW Washington)	606	402	662	342	569	378
District 5 CRDC)	1,931	209	1,915	190	1,791	289
District 6	302	44	292	47	249	78
District 7	204	315	218	303	180	315
District 8	197	22	208	15	196	22
District 9	839	1,228	724	1,415	817	1,203
District 10	395	417	407	402	384	410
District 11	125	69	144	46	134	53
District 12	33	1,157	41	1,154	73	1,114
Miscellaneous	19	8	22	3	23	4
California Provisional	233	61	250	47	197	91
Southern Locals	2,385	160	2,396	149	2,402	136
TOTAL	8,580	12,840	9,061	12,302	8,492	12,554

Source: Official Report of the International Tabulating Committee (Copy in the Adolph Germer papers, University of Oregon Library). Larsen, Laux and Greenall were the left-wing candidates. Mitchell was a B.C. White Bloc leader.

The Cold War

In October 1946, C.E. Wilson, head of General Electric, summarized the situation in the United States by saying: "The problems of the United States can be captiously summed up in two words: Russia abroad, labor at home." World War II had bequeathed a world empire to U.S. capitalism — if only the Soviet Union and its allies around the world would not get in the way — while at home, the stimulus provided by the war had boosted the economy out of its 1930s depression. Now America's post-war prosperity only depended upon the government's continuing purchases of military goods and labor's acceptance of productivity increases. In short, economic prosperity for the United States required a war-time economy during peace time. An "enemy" had to be created and for this role Soviet Communism was eminently suitable.[62]

The ideological "scare" was led by the U.S. Chamber of Commerce which issued a series of pamphlets on communist infiltration of American institutions. One of the pamphlets, "Communists Within the Labor Movement, The Facts and Counter-Measures," suggested that employers could encourage opposition to Communist labor leaders by mailing anti-communist literature recommended by the Chamber to the homes of potential anti-communist leaders. The pamphlet advised employers that church and veterans' groups would help to create a "general atmosphere hostile to communism." It recommended "experienced unionists...Socialists and Social Democrats, and non-Stalinist communist groups" as "the best fighters against Reds in labor."[63]

But as the crusade against communism was being carried to the American people through the news media, the business community prepared what John L. Lewis called "the first ugly, savage thrust of Fascism in America" — the Taft-Hartley Act. Taft-Hartley, passed in 1947, stripped labor of most of the rights it had won with the passage of the Wagner Act. It gave employers the right to enjoin labor from striking, established a sixty-day cooling-off period during which strikes were forbidden, outlawed mass picketing, denied unions the right to contribute to political campaigns and abolished closed shops. Most importantly, however, the law required all union officers to take oaths that they were not members of the Communist Party. Failure to do so disqualified the union involved from recognition by the National Labor Relations Board.

The anti-communist provisions of Taft-Hartley put Communist trade unionists in an untenable position: if they refused to sign, their union's rank and file would be deprived of NLRB services; if they signed the affidavits, they were subject to perjury; if they resigned, the trade union movement's left wing would be effectively severed. Locals whose officers did not sign the affidavits were considered outlaw locals and the CIO chartered parallel locals to organize against them. Eventually, the CIO expelled the unions which had not complied with the Taft-Hartley Act. Furthermore, the Act gave anti-communists within the union movement a new weapon: by charging that an opponent was a "Communist," a candidate for office or representative for a certain policy position could be effectively intimidated and discredited. Union leaders became afraid of being investigated by the government or by the CIO itself. In such a climate, they altered their activity to appear less militant; consequently, the more conservative unionists were able to advance their careers, win elections and move the CIO rapidly to the right.

At the time, James Fadling was IWA president. Fadling, a native of Oklahoma, had been a member of the White Bloc since the early faction fights in the Southwest Washington District. When he was drafted for military service in 1943 he had been First International Vice-President and next in line of succession for the presidency. Claude Ballard, Second International Vice-President, had moved into the position vacated by Fadling. In the fall of 1944 while Fadling was still absent, President Worth Lowery died of a heart attack and Ballard filled the vacancy. When Fadling returned from the Navy in 1945 he claimed the presidency and, after an in-house fight, won it. He won reelection in 1947 and 1949.

On July 22, 1947, the International Executive Council met to consider the Taft-Hartley Act. President Fadling recommended that the union "go on record to comply with the NFRB certification provisions" of the law, but left-wing board members, Karly Larsen, Ilmar Koivunen and Ernie Dalskog all spoke against Fadling's recommendation. Dalskog argued that the IWA should defy the law. "But at the same time," he added, "we must strengthen our organization so that we not only defy it, but defeat [it]...the emphasis should be on defying the bill rather than complying with it."[64]

Many of the conservatives expressed reservations about the blatantly anti-labor nature of the law, but the temptation to use it for their own political interests could not be resisted. Rationalizing their opportunism as mere acquiescence to the status quo, the White Bloc members of the council supported compliance. "The law has been passed. We have it now," argued vice-President Bill Botkin. With only the three left-wing members of the council voting nay, the Council agreed to comply with the terms of the Taft-Hartley Act.

The most extraordinary point to note about the Council's action was that it came at a time when the official National CIO position was non-compliance. Amazingly, even after being informed of the National CIO position, the Executive Council reaffirmed its position at a meeting on August 21. Although the CIO eventually did go on record favoring compliance, it was premature actions like that of the IWA's White Bloc which encouraged it to do so.[65]

On that same August date the Taft-Hartley issue went to the IWA's International Convention in St. Louis.

There was only one Canadian delegate at this convention. All the others had been stopped at the border and, according to Jack Greenall, "all known Communists were refused entry." Greenall, who was already in the U.S. on other business, "was the only known Communist from Canada at the convention."[66]

Two resolutions came out of committee for consideration by the convention delegates. The committee majority recommended a resolution protesting the anti-labor character of Taft-Hartley but resolving that the IWA "comply with the NLRB certification provisions of the Taft-Hartley Law..."[67] The minority recommended a resolution which called upon the union "not to use the facilities of the new Labor Board" and "to resolve all issues between our union and the employers through bona fide collective bargaining and other peaceful means wherever possible."[68]

Although communism was not mentioned in either of the resolutions, it was immediately made the central question in debate. Carl Winn, from the Everett Boom and Rafters Local, spoke in favor of the IWA's compliance with the Act, and stated that:

All they [those speaking against compliance] object to is signing an affidavit that they are not a member of the Communist Party or that they don't advocate the overthrow of the American Government by force or Revolution. I would like to know what some of you people would think and do if the officers of this International union refused to sign that affidavit? I know Carl Winn isn't; he will sign an affidavit that he is not a member of the Communist Party."

Joe Brant, from Aberdeen Local 3-2 and a member of the resolutions committee, said, "I want to state here and now, the only issue before this convention is that one stipulation whether or not your officers should sign that stipulation. Now that is the issue and regardless of what these other speakers say, that is the only one...[69] Finally Vice-President Botkin promised that, "As one of your International officers, I assure you that I have no objection to signing an affidavit that I am not a member of the Communist Party."[70]

The "test vote" came immediately after Botkin's speech on a motion to cut off debate. How the absence of the Canadian Communists affected the debate on Taft-Hartley cannot be known, of course. As it was, the motion to end debate did not get the two-thirds majority it needed to pass, although eventually the minority position was voted down, and IWA compliance with Taft-Hartley was passed.[71]

The anti-communist International leadership quickly took advantage of the Taft-Hartley Law. On September 18, 1947 International President James Fadling wrote to Larsen and Ed Laux, the International's left-wing secretary-treasurer who had been elected with Larsen, asking them to sign the necessary affidavits or "tender your resignation(s) immediately." Larsen and Laux refused to sign, saying they did not want to become "a legal party with Taft-Hartley and the Labor Management Board in destroying industrial unionism in the lumbering industry." Drawing a parallel with fascism in Europe, they said that they Taft-Hartley Act had branded "everyone and every organization that has the courage to fight for the right of the common people, communists, [with] the same formula used in Europe."[72] Larsen was the first CPUSA member to resign under the Taft-Hartley; this gave his case added national and international significance. Joseph Starobin, in *American Communism in Crisis, 1943-1957,* describes Larsen's resignation as a sign that, nationally, the Communist Party's resolve to fight the Taft-Hartley Act was weakening.[73]

The White Bloc also employed the Taft-Hartley Act as an organizing tool against left-wing locals. In several cases, the International set up paper locals to counter those locals whose leaders had not signed anti-communist affidavits. The confused workers were then encouraged to abandon their "communist" local and come over to the "legal" local. During a strike by Bellingham Local 46, the International's secretary-treasurer Carl Winn called the local a "communist organization" and urged the strikers to "go back to work under the promise of a separate charter." In another incident, workers in Gardiner, Oregon were offered "a local charter with a business agent and their own union officials" because their local, 7-140, had not complied with the Taft-Hartley Act.[74]

Taft-Hartley and the B.C. IWA

The Taft-Hartley crisis combined with other issues to strain and finally break the ability of the beleaguered B.C. District to defend itself. Unfortunately for the union, the District's successful 1946 strike had raised employers' combativeness to new levels; with control of the province at stake, lumber bosses had initiated a campaign of espionage and disruption against the IWA. Meanwhile, the CCF and the Canadian Congress of Labour, both bent on ruling the B.C. labour movement, continued to hammer away at the IWA's Communist leaders. The International's White Bloc realized that the opportune time had finally arrived to eliminate the Left in British Columbia once and for all.

Confronted with this complex network of political forces, the B.C. District Council had little choice but to seek the ultimate solution — secession. The events leading up to this "Canadian Breakaway" or "October Revolution" involved conspiracy and deception on a level nearly unparalleled in the North American labor movement. The sequence began on October 8, 1947, when Jack Greenall was asked to resign his position as an International trustee. When he refused to resign, President Fadling suspended him on October 25,[75] thus increasing the tension between the IWA's International office and the B.C. District Council which for years had protested the International's attempt to destabilize the left-wing control of the District. Greenall, in fact, had been at the vortex of the struggle since 1942 when he had become the first organizer fired by the International for political activity.

Fadling's suspension of Greenall prompted the Left to demand a recall election against Fadling and at its March 1948 board meeting, the British Columbia District Council joined the recall movement. When Canadian organizers Tommy MacDonald, Les Urquhart and Mike Freylinger signed the petitions calling for the recall election, they were immediately fired by George Brown, the International's Director of Organizing. Then, even though the petitions had not yet been received,[76] the International's executive board declared them invalid at its March 10 meeting.

By this time, the International White Bloc members had established their own allies within the British Columbia District, and the Canadian White Bloc opened a new attack on the District's leaders, announcing that some district funds were missing. The charge was never proven, but its re-examination is useful for what it reveals about the White Bloc's tactics.

The original charge was made on February 29, 1948 by Local 1-357. Based on a report by one of its members, L.E. Vandale, the Local claimed there was $9,372.85 "not accounted for" in the District's accounts for 1946 and 1947. The District denied the accusation, basing its defense on an audit conducted by Eric Bee of the Trade Union Research Bureau. To clear itself, however, the District ordered a second audit by the firm of Riddell, Stead, Graham and Hutchison. As was the common practice following an audit, copies of the District's payroll vouchers for the years 1946–47 had been destroyed after Bee's December 13, 1947 report which had reconciled the District's books. The Riddell report, consequently, noted that in several places the District accounts were "not substantiated by any supporting evidence other than the cheques drawn on and paid by the bank." District officers pointed out that duplicate vouchers could be located in *local* union files but the auditors did not seek these out before they completed their report. Thus, the second audit produced no evidence that the funds had been misused, but it also failed to offer concrete evidence to absolve the District of the charges against it. To make matters worse, the funds now associated with the missing vouchers totaled about $150,000, and now when the White Bloc began its radio broadcasts attacking the District leaders, this figure was construed as the *missing funds.*[77]

The International Executive Board set up a three-member committee to investigate the financial affairs of the British Columbia District; but the District officers, charging outside interference, refused to cooperate. When the committee released its report, "it accused the district of 'gross mishandling' of funds and 'wholly inadequate bookkeeping,' but it could find no proof of any 'criminal activity.'" However, a June 21, 1948 report by the District's trustees concluded that the original charges made by Local 1-357 were complete fabrications. The report also concluded that aside from the vouchers which had been legitimately destroyed, differences between the two audits could be accounted for by the use of two different auditing techniques.[78]

The entire episode, however, had served the purpose of casting doubt on the integrity of the District leaders. The International now began overt intervention in the District by establishing its own newspaper and radio program, "The Voice of the IWA." In a January 26, 1948 broadcast over radio station CJOR, Al Hartung, International First Vice-President of the IWA, asserted that if Harold Pritchett, Ernie Dalskog and Karly Larsen "are not on the bosses' payroll, then the lumber operators are getting a lot of free help. . ." He asked his radio listeners if it could be "that Joe Stalin feels it would hurt his cause if the workers received more wages and better working conditions." He asked B.C. workers to "remove from office those who are against IWA policy. . . and back Brother Fadling to the limit."[79]

The red-baiting got worse as the year wore on. By August "The Voice of the IWA" proclaimed that "communists should be criticized and exposed." It charged that working conditions for B.C. loggers were "almost paradise compared with conditions in the slave labor camps in communist controlled countries. When the communists get power you don't even have to open your mouth in criticism of them. If they merely suspect that you are not in sympathy with their reign of terror, you disappear. You may be killed or you may be sent to a slave labor camp. There you are expendable. You are fed just enough to keep you alive and working."[80]

The ideas and energy for the International's new anti-left campaign were not generated by the woodworkers themselves. Hartung's radio speeches were written by Ellery Foster, the International's first research director, who had been hired in 1945. Foster was a professional forester and avowed anti-communist who had worked for the War Production Board in Washington just prior to his four-year stint with the IWA.[81]

Adolph Germer also reappeared in the IWA's affairs at this time. Germer had been serving as the CIO's ambassador to the World Federation of Trade Unions and had returned to lobby for the Marshall Plan, the instrument for economic aid which was supposed to provide a Western European bulwark against Communism. Debate in the U.S. trade union circles over its meaning and merit was intense. Germer appeared at the IWA's executive board meeting on March 9, 1948 to tell the board that the Marshall Plan was more than a "food for peace" plan; he warned them that the United States may be forced to send guns to defend Western Europe against "communist aggression." Germer linked the Marshall Plan's anti-communist imperatives with the effect of the Taft-Hartley Act on the domestic scene. "The Communists all over the world, those in the United States included, speak the same language and sing the same songs." Germer concluded by saying, "The choice for guidance in this case is between the CIO on the one side and the Communist Party...on the other. Take your choice."

Board member Ilmar Koivunen argued against support for the Marshall Plan, saying the money might be used for guns and ammunition rather than food. But Germer, as he had so many times before, carried the day and the board voted support for the Marshall Plan.

The involvements of Foster and Germer at this critical juncture represent some historical threads that run throughout the IWA's history. Germer, we have already noted, was a trade union professional who had never worked in the industry and was not an IWA member. He was a representative of a segment of trade union leadership that was only interested in securing its own place in the post-war order. Foster was a maverick intellectual infatuated with the ideas of the 19th century Russian anarchist Kropotkin. In the late 1930s he had vowed to stay away from government for the rest of his life, but in the early 1940s he had found himself on the War Production Board as the lumber expert. It was in that capacity and through an introduction by CIO leader Leo Goodman that Foster met Al Hartung, Carl Winn and Worth Lowery from the IWA. His anti-government and decentralist ideas undoubtedly appealed to their frontier individualism and he was invited to be the IWA's research director.

Both Foster and Germer possessed the writing and speaking skills that individuals like Hartung lacked, and their connections with state agencies and international labor bodies gave them access to useful information and imparted a wider legitimacy to the cause of the White Bloc. Thus they became the mediating agents through which state and employing class interests were able to penetrate and influence the direction of the IWA in the postwar years.

The CCL and CCF

During February, 1947, just when the employers and the International had been launching their attacks on the left-wing IWA leaders in British Columbia, the Canadian Congress of Labour and the Cooperative Commonwealth Federation had renewed their efforts to dislodge the Communist Party from trade union leadership in the province. Their attack was spearheaded by Bill Mahoney, another organizer from the Steelworkers Union, who was sent to British Columbia in November 1947, "to take charge of a two-year campaign to rid the Congress unions in the province of their Communists." Mahoney set three targets for his campaign: the Vancouver Labour Council, the British Columbia Federation of Labour, and the IWA. By January of 1948 all twenty-one seats of the Vancouver Labour Council belonged to Congress-backed anti-communist delegates.[82]

Mahoney next tackled the provincial labor federation; this task required a simultaneous attack on the IWA, which he knew was not going to be a push-over. The 1946 International elections had proved that rank-and-file support for their Communist leaders was strong. Furthermore, the existing anti-communists within the leadership were said to be "basically Trotskyite" and therefore not to the liking of the CCF.

In early March, 1948 the B.C. White Bloc was "completely smashed" in district elections, winning a majority only in the New Westminster local. Mahoney, therefore, used the New Westminster local as his staging area, working closely with Stewart Alsbury and George Mitchell. Together they exploited the tensions created by the International's suspension of Jack Greenall and the dissension over control of the organizing program to build a campaign against the District's leaders. Weekly radio broadcasts were established to publicize the CCL's line on labor controversies in British Columbia. Every effort was made to involve existing anti-communist locals in the political affairs of the provincial labor council, while a moratorium was put on the organization of new locals unless they showed anti-communist potential.[83]

At the B.C. Federation of Labour convention in September, 1948, rightists captured control. Prior to the convention the CCL had suspended the

International Mine, Mill and Smelter Workers' Union, the IWA's major left-wing ally, thus depriving the left of twenty-two delegates at the convention. Even at that, the left carried most of the early convention ballots although it was only by a single vote each time. Mahoney, having calculated his support very carefully, suspected that one of his supporters was double-crossing him, so he sent an aide to sit beside the "suspect" and, as historian Irving Abella has recounted, "Sure enough. . .on the next vote for the position of second vice-president the Congress candidate Alsbury defeated the left-wing candidate Alex McKenzie by the same 66 to 65 vote." In the most important election of the day, White Bloc representative George Home defeated Harold Pritchett by the same narrow margin. When the voting was over, the CCL-CCF delegates had captured five of the nine executive seats. [84]

The efforts of the Canadian Congress of Labour and the leaders of the IWA International had now merged, and by this time, Mahoney had attended an International Executive Board meeting in Portland to discuss strategy and begin joint efforts to pursue the charge that funds had been mishandled by the B.C. District. Late in the summer of 1948, International President Fadling "demanded Congress assistance in 'undermining' the District leadership," and the District heard rumours that the International was planning to seize the District's assets and put it into receivership. On the U.S. side of the border, anti-communists from the Columbia River District Council were mobilized to attend meetings in British Columbia to organize opposition to the B.C. District leaders. [85]

As early as July, the B.C. District warned the International to cease its campaign of disruption or face the possibility of secession, but when the Port Alberni local passed a resolution calling for a referendum on the secession question, Fadling remained insensitive. Canadian Congress of Labour head Pat Conroy called the impending secession "extremely fortunate" since it would save expulsion proceedings.

Additional developments, however, finally forced the District toward secession. After allowing B.C. District leaders to cross the border for a brief period in early 1948, the U.S. government again closed its border to them; apparently, the Canadian Communists' vocal condemnation of the Taft-Hartley Act had caused the change in border policy. And for the second straight year, U.S. border authorities had also denied the *entire* left wing of the Canadian delegation permission to attend the IWA convention in Portland. In addition, the International now appeared to be close to putting the District into receivership. Therefore, on October 3, 1948, by an overwhelming margin, the British Columbia District Council voted to secede from the International Woodworkers of America and form the Woodworkers Industrial Union of Canada (WIUC). [87]

The Woodworkers Industrial Union of Canada, 1948-1951

The secession vote was taken at the regularly scheduled quarterly meeting of the District. Besides the District officers the meeting was attended by International President James Fadling, board members from union locals and seventy-five delegates. The meeting had hardly begun when a floor fight erupted over a resolution condemning employers for raising board rates (the price that loggers are charged for their camp meals) subsequent to the recently settled contract negotiations. The resolution cited the officers of Local 357 and James Fadling for having engaged in divisive activities which had weakened the union's effort to gain a more favorable settlement and had invited the employers' heavy-handed action on board rates.[87]

Fred Wilson from Local 1-80 told the meeting that "the bosses would never have [raised board rates] unless they were sure that they had the full cooperation of our International President, and Alsbury, and some of these guys in New Westminster."[88] The anger of the delegates boiled over when Fadling rose to answer Wilson's charges, and he was greeted with shouts of "Sit down! Sit down!" With considerable effort the chairman quieted the meeting and allowed Fadling to speak. The International President warned the delegates that the resolution was a "two-edged sword" which required them to condemn the bosses and their International leader at the same time, but despite the warning, the delegates did exactly that by passing the resolution.

After that, the gathering inched closer to the disaffiliation question by instructing the District Officers to "take any and all steps which in their discretion they deem necessary to fully protect and preserve assets, funds, and property of the membership at present contained in B.C. District Council No. 1, IWA." Then at last, the debate on disaffiliation began.

In the resolution the officers enumerated the reasons for disaffiliation. These included violation of the District's autonomy by the International; slander of District leaders; appointment of disruptive organizers who were paid with funds collected from the District; signing of the "yellow dog" Taft-Hartley anti-communist affidavits by International leaders; the setting up of underground White Bloc caucuses within the District; the banning of Canadian workers from the International convention by the U.S. Department of Immigration.[89]

Lloyd Whalen was the first to respond to the disaffiliation resolution. He called it "a continuation of...a policy completely alien to the working class..." Following Whalen, Stewart Alsbury accused the leadership of District 1 of having "a political axe to grind" and of trying "to take the woodworkers of B.C. out of the International Union and put them into the

Third International..." The debate intensified when Fred Fieber from Local 1-357 labeled the District officers "a bunch of liars" for the charges they had leveled. Dalskog, chairing the meeting, demanded Fieber retract his statement. "Take it back! Take it back!" shouted the delegates.[90]

Eight speakers rose to support the motion for disaffiliation. Edna Brown spoke for the Ladies Auxiliary and credited the District 1 leaders with being "the first trade union people who have agreed that women are important in the trade union movement."[91] (Later Fadling attempted to dissolve the women's auxiliaries because of their opposition to the Marshall Plan, and International Secretary Carl Winn stated that women were only good "to work in the kitchen and to service the American Navy.")[92]

Dalskog denounced the domination of the trade union movement by those who supported the foreign policy of big business and the U.S. Congress. When Fadling attempted to respond, the following exchange took place:

J. Fadling: Mr. Chairman, as International President, I...

Voices: Sit down. Sit down. The question on the motion. Sit down.

S. Alsbury: Bro. Chairman, give him five minutes.

Voices: Sit down. Sit down. Question. Question.

E. Dalskog: The majority of the delegates seem to want the question put now.

J. Fadling: As I understand, you are refusing me the floor.

Voices: Question. Question. Question.

J. Fadling: You haven't passed this resolution yet, you know. Do you refuse me the floor, Bro. Dalskog?

Following this exchange (Fadling was never given the floor), the motion was read once more and passed. Immediately afterwards, Alsbury took the floor.

S. Alsbury: Bro. Chairman, on behalf of the New Westminster delegation, including the board member, we wish our vote recorded as being in opposition.

E. Dalskog: The motion is carried.

S. Alsbury: Inasmuch as we are no longer members [sic] of the International Woodworkers of America, we will have to leave.
(Loud cheers and applause.)

Voices: Take Bro. Fadling with you.
(At this point a number of delegates left the meeting.)[93]

The remainder of that day's meeting dealt with administrative details. The ex-B.C. District officers were made temporary officers of the WIUC, October 23 was set for the first WIUC convention, and temporary by-laws were passed. Finally, a unanimous standing vote for a "Declaration of Independence" launched the "October Revolution."[94]

The leaders of the WIUC viewed its first convention as a sequel to the 1937 IWA break with the United Brotherhood of Carpenters and Joiners (UBCJ).[95] "We are living in a period of change," said Harold Pritchett to the first WIUC convention, "and anybody who flies in the face of change is either unintelligent or has a purpose for [it]." Pritchett drew parallels between Bill Hutcheson, the reactionary leader of the Carpenters Union in 1937, and Fadling. He pointed out that conservatives had opposed the movement to go CIO in 1937 and were presently leading the opposition to the WIUC. But, said Pritchett, "in spite of what Fadling [said] in 1937, and in spite of what Fadling said in 1948, and in spite of the bosses, the Woodworkers' Industrial Union is here to stay and to grow and to represent the best interests of woodworkers across the length and breadth of this great country."[96]

The document governing the WIUC reflected the members' respect for democratic organization. Among its provisions were those setting officers' salaries at $65 per week with increases tied to wage gains in the industry, providing for referendum election of all officers, allowing referendum recall proceedings to be initiated by twenty percent of the members and a simple majority vote needed for recall, electing job stewards at the local level, and banning discrimination by reason of race, color, sex, religion or political belief.[97] Ernie Dalskog summed up the spirit in which the new constitution was adopted and called attention to the national scope of WIUC plans:

> We don't want an organization where they have a business agent that handles everything; we don't want to have an organization that is run from the top — we want to have an organization in which the membership is the determining factor. That is the type of organization that the Woodworkers' Industrial Union of Canada is going to back and that is the kind of an organization that will extend not only to the Province of British Columbia, but into all of the other Provinces of Canada in the very near future.[98]

Bert Melsness called it a "great day for the woodworkers in British Columbia and Canada." Alex Shouldra from Local 71 predicted a membership increase from twenty-seven to forty thousand in British Columbia. Thomas McDonald from Local 423 said the IWA would go down in history "as one of the greatest unions... but that is going to be only a shadow of what we will do under the Woodworkers' Industrial Union of Canada."[99]

The WIUC called for all woodworkers to revoke their IWA dues check-off and begin paying dues to the new organization. In operations where a majority of workers had joined, applications were to be made for certification. The key to WIUC success would lie with the job stewards and their aggressive pursuit of grievances. [100]

But although convention delegates were enthusiastic, ominous clouds lay on the horizon. The Canadian Congress of Labour rejected proposals for affiliation; companies with whom the IWA had held contracts refused to recognize the WIUC as the bargaining agent; organizers reported confusion and defections among the rank and file. Most importantly, the employers were doing everything they could to wreck the WIUC. The new union's organizers cited numerous instances of employers attempting to harass them and intimidate WIUC members. [101]

Despite its legacy of struggle against company unionism, the IWA allied itself with the employers and the latter quickly availed themselves of the service. On October 9, 1948 British Columbia Forest Products Limited announced that its contracts with the IWA remained in full force and that it "would recognize and deal only with [the IWA]." When entering company property, IWA organizers were granted special privileges by the employers including escorts by "superintendents and foremen and even the manager." In one case "IWA administrator Claude Ballard and Lloyd Whalen were granted permission to bring their sound truck onto company property and space was cleared in the yard, by the company, for the meeting." In late October radio station CJOR allowed the IWA to take over the District's weekly radio program, "Green Gold," and to broadcast under that name "in the same time slot using the same sound effects" as had the District. [102]

The International appointed anti-communist leaders to the British Columbia District offices to enable the IWA to continue its union activity even though most of the membership had in fact disaffiliated. The Left charged that these "provisional" locals and District Council apparatus were in reality only "paper organizations" designed to disrupt a legitimate organization, the WIUC.

One individual taking office in a "paper local," Joe Morris, eventually rose to the presidencies of the Canadian Labour Congress (CLC) and the International Labor Organization (ILO). Morris recalled "[moving] into the office that had been vacated by the WIUC people [and starting] with a borrowed typewriter, a borrowed mimeograph machine, an old table, a chair and a couple of packing cases. We had no records and no money." [103] But the International quickly moved to wipe the slate of debts owed by the "paper locals" to the International during the fight with the WIUC. They also cancelled per capita dues, and the cost of dues stamps, office supplies and debts incurred by the locals prior to the breakaway.[104]

The Battle of Iron River

The fight between the IWA and WIUC turned violent at Iron River, a logging camp and WIUC stronghold south of Campbell River on Vancouver Island. In early November MacMillan fired two loggers, Anton Johnson and George Nichols "on the excuse that they were not cutting enough timber." When the crew struck to enforce the seniority rights of the fired loggers, the IWA, represented by Alsbury, Whalen and Tom Bradley, attempted to settle the dispute by negotiating directly with the employer and encouraging the crew to return to work.[105] As a consequence, on November 15 one hundred WIUC loggers struck the entire MacMillan operation. Upon the IWA's request, the Labour Relations Board declared the strike illegal, and the International moved in to break the strike. Then on December 8 the IWA and the fledgling WIUC met in what the WIUC paper labeled "one of the most outstanding and shameful incidents in Canadian labor history."[106]

"We very carefully laid our plans and developed our strategy," recalled Joe Morris of the IWA, "and decided that we would escort our people to work..." Strikebreakers shepherded by IWA representatives arrived before daylight and met the WIUC workers in the glare of automobile headlights. "When we drove into camp with our people, [communists] stormed across the road and into the camp," said Morris. "The provisional president of the Regional Council [Stewart Alsbury] was kicked so badly that we had to take him [and two others] to the hospital. However, we were successful in getting our people in the crew-cars and out to the job..."[107] In fact, only twenty-five of their people went to work that day.

That night a force of 150 "thugs," few of whom were loggers and many of whom were reportedly from Oregon and Washington, were bussed into the district. "This mob," according to a *Canadian Woodworker* account, "was escorted to the picket line...by 26 provincial police...Unable to provoke a fight with the token picket line of 14 men and three women facing them, they burned down the lean-to shelter used by strikers."[108] For the second straight day only twenty-five scabs went to work.

Five Iron River strikers were tried on charges of having assaulted Alsbury and Tom Bradley. The WIUC defendants proudly turned the courtroom into a political forum. Mike Farkas took the stand and testified that "he had beat up Alsbury and probably broken [Alsbury's] ribs when he threw him over his shoulder to the ground because Alsbury had tried to lead scabs through a picket line." A defense witness, Danny Holt testified "he had seen Farkas beat up Alsbury until he yelled for mercy. . ." Holt testified: "I told Farkas, 'Hit him again, Mike,' and Mike did."[109]

The trial and its aftermath revealed further details of the IWA's conspiracy to disrupt the left-wing WIUC movement. Alsbury admitted that the company had "invited" him to herd scabs to the Iron River site. The use of strikebreakers from the U.S. had been planned as early as the IWA convention in October when a call had gone out for volunteers and White Bloc loggers to infiltrate the Vancouver Island operations. Evidence also indicated that the right-wing provisional leaders had used District-owned strike funds to finance the scabbing operation at Iron River.[110]

Both sides claimed victory at Iron River, but the new union, in its pursuit of independent unionism for Canadian woodworkers, continued to organize in the mills and woods, while fighting through legal channels for its right to exist. Newsletters hammered at the worsening conditions of employment and the rapidly accumulating unresolved grievances. They attacked the IWA leadership for loyalty to the companies, and for the expenditure in the United States of dues money collected in British Columbia; then as details of the International's disruption campaign in B.C. became available, they were also reported. At first, these newsletters came from Vancouver, New Westminster, Prince George, Cranbrook and Victoria, but later the WUIC began a province-wide newspaper; initially it was called the *Canadian Woodworker* but renamed the *Union Woodworker* in March of 1949.[111]

The courts provided the WIUC with additional challenges.

By mid-December of 1948 locals in Cranbrook, Nanaimo, Courtenay and Port Alberni had applied to the Provincial Labour Relations Board for certification in fourteen operations. By mid-February the applications numbered 34, yet only one election, in Victoria, had been held. The IWA won that election by one vote. Following the Victoria election, the IWA asked the courts for an injunction to half Labour Relations Board action on other WIUC certification requests. That move was blocked by reciprocal WIUC court action, and in early April, the Labour Board moved on the WIUC applications. The progress was slow, however, and by the end of May, 1949, the WUIC was certified in only nine operations.[112]

By the second WIUC convention it was becoming clear that disaffiliation from an International union was easy to declare, but difficult to consolidate. The convention, held April 2 and 3, 1949, called for the "re-uniting of woodworkers' ranks" based on "rank and file control and full trade union democracy." This provoked the IWA's International Newspaper, reporting on the WIUC convention, to claim Pritchett was seeking reaffiliation with the IWA.[113]

On April 17 the strike at Iron River, which had been maintained through the winter months, was called off. Defeat in this first major strike conducted by the WIUC was doubly demoralizing to the union ranks because only one week earlier, Vice-President Ernie Dalskog had been jailed for contempt of court. Dalskog had refused to obey a court order that he turn over to the IWA $130,000 in strike funds which had been entrusted to him at the disaffiliation convention. Now, the IWA International and the Provisional District Council claimed the money was rightfully theirs and that Dalskog and the "reds" had stolen it. Vancouver newspapers headlined the "theft" for weeks. When imprisonment failed to change Dalskog's mind, the court attacked the entire WIUC leadership; Harold Pritchett, Jack Forbes, Hjalmar Bergren and Bert Melsness were all ordered to appear in court. The WIUC's principal organizers were thus bound to a court docket when they should have been in the field organizing.

The WIUC never relinquished its claim to the strike funds. Yet the reality of the situation was that by this time the IWA had won the jurisdictional battle and actually represented the bulk of British Columbia Woodworkers. Furthermore, the IWA was about to enter into contract negotiations with employers, and, if the WIUC was to act in the interest of the rank and file, it had no choice but to place the strike funds at the IWA's disposal. In early May the $130,000 strike fund was handed over to the court and subsequently to the IWA. Dalskog was released after spending nearly a month in jail.

A second step toward unity was taken in June when the WIUC National Council voted unanimously in favor of supporting IWA members "in any action that may be necessary – up to and including strike action – to win the demands laid down for the 1949-50 contract." The *Union Woodworker* editorialized on the need for unity and proposed a "shoulder to shoulder fight" against the deteriorating working conditions in the industry.[114]

And reporting on the expulsion of the United Electrical Workers Union and the Mine, Mill and Smelter Workers Union from the CIO, the editor pointed out, "a year ago, the main bulk of the labor movement was more or less united in a common front. Today, it has been driven asunder, transformed into warring factions, with the remnants of the CIO raiding and wrecking in obedience to Murray's policy of 'divide and conquer.' A united CIO movement no longer exists."[115]

New problems arose for the WIUC when employers refused to accept revocations of IWA dues-checkoffs submitted by WIUC supporters unless they were signed in the presence of company officials – a policy designed to intimidate the workers. The Labour Board, however, condoned the procedure and eventually began siding openly with the IWA and employers. In the case of the Columbia Contracting Company's planer operations, the Board granted the IWA certification without even holding an election.[116]

With the employers, the State and the international labor movement allied against the WIUC, the drive toward unity was accelerated. By late 1949 polemics against the IWA had subsided and in May of 1950, the *Union Woodworker* endorsed an IWA proposal for a union shop. In July the paper stated:

> ...a decisive turn has now to be made in our work and ground prepared for the next step forward. There is much to be done and done quickly.

> Today, trade unionism is at a lower ebb in the lumber industry than it has been for years. In many operations, no union of any kind exists, although the workers have amply demonstrated a willingness and readiness to fight. In other operations, one or both unions exist in name only. Only in a few places is there actual organization on the job. This situation must be remedied without delay.

> It is with this situation before us that we must face the issue of how to unite and rally the workers to fight for their immediate needs and for the future.[117]

In August 1950, the WIUC officers advised woodworkers to "join hands together as workers having common problems by building one powerful industrial union for the lumber industry," and with only a few more parting shots at the IWA leadership, the Woodworkers Industrial Union of Canada was abandoned.[118]

Slowly, all the WIUC locals drifted back to the IWA, the last being Local 405 in Cranbrook. By the mid-1950s the only noticeable effect of the breakaway was that the Communists who had led it were now isolated. In the end, the prophecy of CCL leader Pat Conroy — that the "October Revolution" might actually be a blessing in disguise for the anti-communists — proved to be correct.

The demise of the WIUC was an inglorious end to one of the most tenacious attempts to build progressive unionism in North America. Pockets of Communist-led woodworkers held out in the U.S. and Canada well into the 1950s, and for years the various anti-communist factions would bludgeon each other with the "Communist" epithet whenever it was convenient. But the fight was over; what followed was only a mopping-up exercise.

The conventional wisdom on the breakaway period is that the secession was a mistake, that the left-wing leadership had "failed" at a decisive moment, and that the rank and file had abandoned the Communists.[119] It is difficult to see, however, that the District's leaders had any choice; had they waited for the trap to close tighter, their defeat would have been even more certain. It is also difficult to concur that the rank and file abandoned its leaders. On the contrary, for at least six years prior to the breakaway, and in spite of unprecedentedly vigorous and resourceful opposition, referendum elections had returned Communist leadersip to District offices. Indeed, White Bloc candidates had been "smashed" in an election only months prior to the breakaway. All the evidence supports the conclusion that the rank and file did stand behind its elected leaders and that the anti-communist movement developed because of the need to sever the strong allegiance between the leadership and the union ranks. This allegiance was attacked by making structural changes in the union (such as breaking up large left-led locals), by using the power of the State against the Left and finally by vicious red-baiting.

The role of factionalism among unionists was also extremely important in deciding this issue. Even in its most unified form, the working class of Canada and the United States faced enormous opposition in the post-World War II period. From shop floor levels to the arenas of international politics, capital had embarked on a campaign to smash labor militancy and solidarity. Capital's resources were enormous when coordinated by the apparatus of the State in the major capitalist countries. In the face of such tremendous odds, to preserve the gains made during the 1930s, the working classes needed maximum solidarity.

Instead, the cold war provided an opportunity for some union leaders to advance themselves by joining the anti-communist crusade promulgated by the capitalist class. By helping to divide the workers from their left-wing leaders and offering management the lesser evil of business unionism, socialist and social democratic activists were able to secure their own futures in the labor movement. What is notable about the period, therefore, is less the failure of the people who led the WIUC movement than the power and resourcefulness of the State and the refusal of non-communist labor groups to accede to the leaders elected by the rank and file.

6
The Post-War Consolidation
of the White Bloc

The demise of the Woodworkers Industrial Union of Canada (WIUC) was a major setback for the left-wing movement within the IWA. The WIUC leaders had built the IWA in British Columbia and sustained it through nearly a decade of attacks by employers, the State and rightist elements within the union itself, and before the WIUC secession, the British Columbia District had accounted for nearly fifty percent of the left-wing vote in International elections. Now that the woodworkers in British Columbia had rejoined the IWA, it was the White Bloc, not the former radical leaders, who dominated the International. World War II had also brought changes to the industry which further undermined Communist influence in the union.

The War and the Wood Industry

By 1943, lumber had surpassed steel and non-ferrous metal as the foremost material need of the armed forces. Staggering quantities of lumber were required: 40,000 board feet for the construction of each C-62 cargo plane; 140,000 board feet for a carrier flight deck; 350,000 board feet for each liberty ship; 9,000,000 board feet per year for ammunition crates and, in 1942 alone, 25,000,000 board feet for portable road beds and bridges. P.T. boats were made from lumber and trainer planes used plywood skins. Wood even replaced some metal tank parts. Warehouses, aircraft hangars and barracks needed lumber for their construction. Lumber production soared to over 36 billion board feet in 1942 with ninety per cent of that going to the war effort.[1]

When large numbers of woodworkers went to war, nearly 50,000 women entered the industry. By 1943 eleven percent of those involved in lumber production were women, compared to one percent in the industry in 1939. Special provisions were made to accommodate the entrance of women into the industry, but most of these changes, such as the separate lunchroom facilities with upholstered lounges at Weyerhaeuser's Longview, Washington operation, were superficial. More often, women entered jobs traditionally held by men — including setting chokers in the woods — with little or no special consideration. Women woodworkers on the West Coast also played an important role in assisting the Aircraft Warning Service in maintaining a 24-hour watch for Japanese airplanes. Members of the Shelton, Washington Local 38 donated their free time to the Service, and the *Woodworker* reported that the family of Gladys White had donated the "entire day of Christmas for this patriotic work."[2]

Although women did "men's work" they did not receive "men's pay." An Illinois survey found that women in the lumber industry were paid only seventy-five percent of men's wages. Discrimination in pay and poor working conditions at the Westfir, Oregon local bought Willamette Valley women into the IWA for the first time in the spring of 1942. A year later Local 5-246 in Springfield, Oregon boasted the first woman officer when Linda McDivit was elected over fellow shingle weaver George Templer. In Mass, Michigan women protested attempts by the Von Platen Fox Company to pay them 14 cents an hour less than the minimum scale specified in the union contract; when the women protested they were promptly fired, but with the help of the International, they won their case in arbitration and were rehired.

But the union went beyond the question of equal pay and took up other issues raised by the war-time employment of women. It called for the establishment of daycare facilities, training programs, and "a fifteen minute rest period in each half day shift for all working people, either male or female."[3]

The union made at least one significant economic gain during the war. In February, 1942, one thousand loggers at the Lamm Lumber Company in Oregon walked out rather than drive their own cars to and from the logging camps which were situated up to seventy miles from town. The rough logging roads were chewing up their tires at a time when the rubber shortage created by the war made it nearly impossible to replace them. The company first offered to provide transportation at twenty-five cents a day per logger, but the crews stayed out until the company agreed to absorb the entire expense. This was a substantial concession, because the industry was moving deeper and deeper into the woods in pursuit of timber and the days when logs and loggers could be moved on rivers or on company rail systems was nearly over. After the war, truck transport would increasingly replace railroads, and the precedent of company-provided transportation which had been set at Lamm would be recognized in union contracts into the 1980s. In addition the IWA was the first union during the war to win a "portal to portal" pay agreement for above-ground workers for travel time to and from the jobsite.[4]

IWA gains are especially significant because the war gave industry owners special opportunities to blunt the militancy of woodworkers. With the critical need for lumber, companies attempted to pay the lowest wage possible, and they met union objections with charges that the union was obstructing the war effort. Operators set a pattern of closing down operations rather than offering a wage that would attract and hold experienced workers. But in late 1942, with over half the logging camps in northern Washington closed, the IWA turned the tables and charged owners "with collusion to undermine the war effort."[5] The *Woodworker* saw the owners' policies as part of the larger pattern of prewar collaboration between the U.S. capitalists and German industrialists, but the immediate objectives of the lumber operators were, in fact, less global.

By refusing to pay competitive wages, the operators had created an artificial labor shortage, and then with this self-made crisis as a pretext, they attempted to bring Federal troops to work in the woods.[6] In the spring of 1943, W.B. Greeley of the West Coast Lumbermen's Association (a business group) and Fred Brundage, the Western Log and Lumber Administrator for the War Production Board, proposed that U.S. Army troops work in the Northwest woods to solve the "labor shortage." Greeley reported that "the Corps of Engineers is in a position to assign its own troops wherever they are needed to increase the production material...[and] that they could be assigned, in suitable military units, to West Coast Logging Camps or sawmills if operators believe that is necessary."

The IWA immediately recognized this plan as a revival of the Loyal Legion of Loggers and Lumbermen (4-L), the U.S. Army's famous "Spruce Division" which had entered the woods during World War I to smash the IWW. IWA President Worth Lowery wrote to the War Production Board that "certain operators in the Pacific Northwest have deliberately sabotaged the lumber production so that troops would be used in the industry and the labor unions destroyed..."[7] As a result of the IWA's response or other factors, no troops were used except in Alaska, where the Corps of Engineers operated its own mills. Nevertheless, the memory of the massive police and military action that had been provoked by the IWW's World War I lumber strike hung over every militant action considered by the International Woodworkers of America during World War II.

Unable to use U.S. Army troops in the woods, lumber bosses developed still another scheme: the use of German war prisoners. The Crossett (Arkansas) Lumber Company "obtained permission (from the War Manpower Commission) to use German war prisoners, claiming they were unable to obtain sufficient manpower to conduct their logging business." About the same time, the Arkansas Security Division of the U.S. Department of Labor recruited 250 workers from Prescott and Paragould, Arkansas and transported them to Washington State for employment. Once again the IWA protested. As a direct consequence, the War Manpower Commission refused all further permits for the employment of prisoners of war in the lumber industry.'[8]

As experienced lumber workers were drafted and as operations were shut down, turnover in the industry workforce rose. This meant that workers who had learned their union skills during the hard years of organizing in the 1930s were replaced by workers for whom the union was a "given". Not having participated in the struggle to obtain a union, they were more inclined to take it for granted. Moreover, by the end of the war, when they had become more experienced as unionists, they were already leaving the industry in large numbers. Working conditions did not encourage long-term employment either. A survey of eighty-seven industries in 1943 showed that sawmills had the highest injury-frequency rate: 58.4 injuries per million employee hours compared to 9.7 for the aircraft industry and 31.5 for shipbuilding. At the end of the war, the wood products industry suffered the second highest "quit rate" of all industries.[9]

Other detrimental war-time effects were more structural. In the IWA's Northern Washington District which had always been a left-wing stronghold, lumber operators had closed mills rather than pay wages competitive with the wartime shipbuilding and aircraft industries in Seattle. At the same time, they opened new mills only a few hundred miles to the south around Portland, Oregon. For historical reasons that were examined in previous chapters, those few hundred miles made a big political difference. Between 1941 and 1951 the Northern Washington District of the IWA lost over one thousand members because of mill closures while the conservative Columbia River District Council (CRDC) gained 7,276 members (a ninety-percent increase) as new mills opened in Oregon. The reasons given for the closures of the northern Washington mills varied but most owners cited the area's declining timber resources. Radical unionists, on the other hand, claimed that the operators were merely carrying out their threat to leave if left-wing union activity continued.

There was also a shift to a less centralized mode of production with a net decline in the number of mills employing over 100 workers and a large net increase in those employing fewer than 100 workers.[10] The large mills which had provided the most favorable conditions for organizing had been hit the hardest by shutdowns. Automobile transportation, moreover, had made commuters out of loggers who had previously been bonded together by life in the logging camps. Finally, a booming housing market and the general affluence that characterized post-war America promised job security and high wages for the woodworkers who had retained their jobs. In brief, the social and economic conditions that made for successful organizing and radicalization of woodworkers in the 1930s were no longer present by the 1950s.

The political effect of these changed conditions was amplified by the organizational changes which were taking place within the IWA. The merger of the weakened Northern Washington District Two with the Southern Washington District Three diminished what remained of the left wing's presence under White Bloc control. The election of Al Hartung as International President, the Seattle Smith Act Trial and the restructuring of the International engineered by the CIO completed the rout of the Communists.

The Merger of Districts Two and Three

Organizationally, the IWA in the Pacific Northwest had been very unbalanced since its formation in 1937. There were three large district councils: Northern Washington Council Two, Southwest Washington Council Three, and the Columbia River District Council Five (CRDC). The CRDC was the largest and most conservative, Northern Washington the most radical. There were seven small district councils: Klamath Basin Council Six; Coos Bay Council Seven; Plywood, Box Shook and Door Council Nine; Eastern Washington and Idaho Council Ten; Boommen and Rafters Council Eleven; and Northern California Council Thirteen. There was also a "Council" of miscellaneous locals scattered throughout Wyoming, Colorado and Montana.

This organizational fragmentation had both strengths and weaknesses. It encouraged community and even ethnic identity in some smaller districts and fostered a high degree of local participation. On the other hand, there were numerous jurisdictional problems that impaired unity against employers. With the emergence of business unionism in the 1950s, the administrative problems of the fragmented district structure overshadowed their value for encouraging local activism.

Most of the administrative problems were caused by Districts Nine and

Eleven, old craft districts inherited from the AFL. Their locals were scattered throughout the geographic jurisdictions of the other districts; therefore, any plan to restructure the region would break up these two districts. This engendered disagreements over how the spoils would be divided among the other districts, the main concern being that the CRDC might be enlarged and become even more dominant. The left-wing District Two feared such a development for political reasons, while conservative District Three in Southern Washington was concerned for the career ambitions of its leaders. Out of these fears came the plan to merge Districts Two and Three, thus providing an eventual counterweight to the CRDC, resolving the problems associated with craft Districts Nine and Eleven, and opening the entire region to restructuring.

A merger between Districts Two and Three had been first considered in 1943, but at the time the left-wing Northern Washington District Two had been large enough to dominate the new formation. For District Three, merger under those conditions would have meant subjugation to "commie policies." In 1945 a merger was discussed again but that proposal, which clearly favored the conservative District Three, was defeated by District Two.[11] The 1948 convention established a committee to work out the jurisdictional problems among districts in the Pacific Northwest, and although it failed to work out an agreement, it laid the groundwork for the 1953 merger of Districts Two and Three.

The Jurisdictional Committee's majority report proposed to the 1949 convention:

> (1) To establish one District Council in Western Washington, which will be comprised of all local Unions in Districts 2 and 3, and those locals in Districts 5, 9 and 11 that are located in Western Washington, west of the Cascade Range.
> (2) To establish one District Council in Western Oregon which will be comprised of all Local Unions in District 7 and those Locals in Districts 5, 9 and 11 that are located in Western Oregon west of the Cascade Range.

The committee's minority report, labeled a CRDC resolution, moved

> That the local Unions now affiliated with the two established Subdivisions, may retain their identity as Local Unions in conformity with the present constitution, but be required to affiliate with the District Council having jurisdiction.[12]

In effect the committee's majority report took four locals with 3,000 members out of the CRDC and placed them in Western Washington. The big Longview local, with its 1,400 members, would have been included in the shift. The absorption of the two subdivisions would have added 18 more locals and 4,700 members to Western Washington while the CRDC would have gained only six locals with 800 members. Had the minority report been followed, on the other hand, the CRDC would have kept all the locals presently under its jurisdiction, and in addition, it would have gained several locals from Districts 9 and 11 which were located in southwestern Washington.

The minority report was defeated in a roll-call vote by 53,125 to 15,416, with the CRDC standing alone in the defeat. When debate was resumed on the committee's recommendation, the CRDC moved to refer the matter back to the committee amid charges that they were trying to stall. The debate on the question of restructuring lasted nearly as long as that on the Taft-Hartley Act, but ended with the convention passing the committee's report. The recommendations were never carried out, however. The committee reported to the 1950 convention that Districts 5 (the CRDC) and 6 (Klamath Basin in southern Oregon) had not agreed to the establishment of Western Washington and Western Oregon districts. The committee was, therefore, dismissed without debate, killing the restructuring effort until 1953.

Exactly what precipitated the final merger of Districts 2 and 3 is unclear. Since the early 1940s, District 2 had consistently supported the merger, while District 3 had wavered, but the latter may have been swayed by the Taft-Hartley Act which, by excluding District Two's communists from office, effectively gave District 3's leadership control over the combined districts. Still another factor was Jim Fadling, the White Bloc leader from District 3 who had been International President since 1945; in early 1952 he had been defeated in his bid for re-election and thus became available for leadership in the new District 23. His election loss gave Fadling a personal incentive to support the merger because he must have seen it as the only way he could regain a powerful position within the International. A third factor contributing to the merger was the fact that a compromise had been worked out whereby Fadling would become president and Karly Larsen vice-president of the new district. Both Districts 2 and 3 would thus be represented in top echelon positions, as would both right and left political tendencies. With this coalition as the glue, District 23 was formed in the early months of 1953.

The coalition did not last long. Al Hartung, who had replaced Fadling as International President, had been hand-picked by the CIO to continue the fight against the Left. A few months after he assumed his duties as head of the International, Hartung dismissed Karly Larsen from office when Larsen went on trial for allegedly violating the Smith Act.

Hartung's Controversial Election

Al Hartung had wanted to be International President since the IWA had been formed in 1937. He had run for the office twice and lost both times, and had to content himself with serving in a variety of important but lesser positions instead. However, by the 1947 convention Fadling had committed several political and social blunders, making it difficult for the anti-communists to consolidate their hold on the union. At one point Fadling had embarrassed the union by settling a strike for a one-cent increase, earning him the title "One-Cent Fadling." And once, when he was supposed to escort General Eisenhower at a CIO convention, "he was so drunk he couldn't stand up," provoking Hartung to complain that the IWA was "the laughing stock of the CIO."[13]

At the same time the left wing, on the strength of its organizing gains in British Columbia, had returned two of its people to International offices in the 1946 elections. Some of the anti-communist stalwarts feared that the entire International leadership would slip from their hands unless they could find a stronger leader for the anti-communist cause.

Consequently, at the 1947 convention the dissidents asked both Hartung and CIO organizer George Brown to run against Fadling for president of the IWA. Fadling convinced Brown not to run. Allan Haywood, assistant CIO director of organization, pressured Hartung into running instead for first vice-president.[14] In his memoirs, Hartung recalled that Haywood called him at the International convention in St. Louis. He wrote:

> No doubt, my friend Germer had called him for help. Mr. Haywood said, "Al, I think that you should consider running for Vice President. As you know, the CIO has invested a lot of money and personal help and it's [CIO President] Phil Murray's and my opinion that unless you get back into the IWA and help straighten it out there will be no IWA before the year is out." I told him that I knew we were in trouble, but I was at a loss of why I should run for the First Vice President. But as I have already stated, they were trying to stop scandal from coming out to the public. He further stated that if I would agree that he would hold my position open.[15]

Hartung and Ernie Dalskog from British Columbia were nominated for first vice-president and Hartung was elected.[16]

When the 1951 Denver convention came along, Hartung was again prevailed upon to run against Fadling, even though ten of the fourteen districts at the convention supported Fadling. The results of the election referendum hung in doubt for weeks and were finally decided by the Executive Board which convened on December 17.[17] For four days it considered challenged ballots. Hartung recalls, "Fadling's supporters kept challenging the committee results. They were crying that the workers were being denied their vote because of some of the rules. The Fadling supporters had the majority on the Executive Board so they kept on making their own rules."[18] After a Christmas recess, the Board reconvened on January 4, 1952, for five more days.

The Board's work appears to have been a highly partisan affair. Ron Roley, Carl Winn, O.D. Armstrong, and John Azevedo carried on the fight for Hartung who was hospitalized.[19] According to Hartung, it was "the Commies" who were carrying on the fight for Fadling, thinking "they could use him by getting him drunk." However, while there were political overtones to this struggle, it is likely that the personal career ambitions of Hartung and Fadling were the central issue. That the left seemed to prefer Fadling is probably attributable to Hartung's greater sophistication in neutralizing Communists in the Cold War period. In the end, Hartung was declared the winner by 19,949 to 18,549.[20] Fadling left the International office and a year later took over as president of the new District 23.

Karly Larsen and the Seattle Smith Act Trial

In 1952, the balance of political forces within the International rested, in large part, on which direction District 23 would ultimately go. The new district was a merger of the Southern Washington District 3, which Fadling and the conservative forces controlled, and Northern Washington District 2 which Karly Larsen and the left-wing forces controlled. If the new District 23 could be consolidated in the anti-communist camp, the fight for control of the International Woodworkers of America would finally be over. If, on the other hand, District 23 were to be consolidated in the leftist camp, the International would remain split and the fight for control would continue.

Larsen was the key individual in the strategy of the left wing, where he was known as "the undisputed leader of all the progressive forces within the IWA." He was the most visible and highest ranking of all the Communist Party members within the IWA, and during the period when the Party was forced underground Larsen had remained the "above ground" leader for the entire Pacific Northwest.[21]

His importance was not missed by the IWA's White Bloc. "As long as Karly Larsen pulls the strings," recorded the minutes of a 1949 White Bloc meeting, "we will have the same commie situation."[22]

The FBI also considered Larsen important; they had begun building a file on him as early as 1941. In a 1945 memo the Bureau called him the "behind the scenes boss of [the] 12th District CP." Larsen, said the memo, was a "powerful figure in Washington State because of [his] key position [in the] IWA."[23] The FBI's campaign against him included total surveillance of his day-to-day activities, planting stories about him in the newspapers and "bugging" his meeting rooms.[24] He was the subject of constant harassment. In 1935 the windows of his home in Lyman, Washington were shot out twice; in 1941 his home in Stanwood, Washington was burned to the ground.[25]

Larsen had been elected International Vice-President in 1947, but resigned within a few months rather than sign the anti-communist affidavits required by the Taft-Hartley Act. Subsequently, he had been re-elected president of the Northern Washington District and at that point signed the affidavits.

In the fall of 1952, while he was attending a union convention in Portland, Larsen was arrested. He and six other Oregon and Washington Communist leaders were charged under the Smith Act and tried in Seattle the following year in the "second round" of the Smith Act trials[26] which were intended to determine whether the defendants had taught or advocated "the overthrow of the United States Government by force and violence."[27]

The six-month-long trial revealed the details of FBI infiltration and information-gathering within the IWA. Two IWA members took the stand and identified themselves as informants.[28] Harley Mores, from Sultan, Washington, testified that he had joined the Communist Party in 1935 and left it in 1937. After the FBI contacted him in 1942, Mores rejoined the Communist Party and remained a member up to and including the time he testified at Larsen's trial. He had, in fact, attended a high level party meeting only days before his appearance in court as a state witness.[29] Mores was an IWA member throughout this entire period and regularly made reports to the FBI on the activities of alleged communists in the IWA, including Karly Larsen.[30] Mores identified Larsen as a member of the Communist Party who had been in attendance at several high level party meetings during the years 1946 to 1951.

Another prosecution witness was IWA member Cecil Maroni, a plywood worker from Bellingham, Washington. Maroni testified that Larsen had tried to recruit him into the Party at an IWA district convention at Olympia on July 29, 1950, and again at the IWA International convention at Minneapolis in September 1950.[31] Maroni also testified that during the years 1943 and 1944 he had made reports to the FBI and to city police on "anything that might be suspicious that occurred in the mill." He had been in contact with the FBI again in 1952.[32]

Larsen's defence was conducted separately from that of the other six defendants, as his Defence Attorney John F. Walthew argued that Larsen's beliefs did not "jibe with the views of anybody in this case." Walthew told the court that, while "communists, to some people, means the overthrow of the government, the evidence will show that Karly Larsen does not believe in that type of communism."[33]

In court, Larsen denied that he had ever advocated or conspired to over-throw the government by force; he also contended that he had joined the Communist Party in 1933, but that he had left it in 1946. His motives for this statement are unclear. In denying party membership after 1946, he may have been trying to protect the integrity of his 1946 election to the Inter-national vice-presidency. Even though he had resigned the position in 1947, to have admitted Party membership during the time he held the post would have increased the pressure on him in his current position as vice-president of the IWA's Western Washington District. Since the IWA was one of the few CIO unions where the Communist Party still had some influence, it was essential to the left wing that Larsen keep his position in that district. The stakes were high enough, therefore, to warrant a defence strategy that would set Larsen off from the other Party leaders on trial. Prosecution attorney Tracy Griffin recognized these facts in his closing argument when he said:

> Apparently the over-all strategy from the very beginning was to spring Larsen. They conceived this whimsy that Larsen, a valuable asset holding the union in the palm of his hand, left the party in 1946. If they could put that curve over the plate, Larsen would be in the clear.[34]

If that had been the defence's strategy, it failed, for although Larsen was acquitted, he was unable to save his job. Several days after Walthew stated in court that Larsen had left the Communist Party in 1946, Al Drawsky, president of Local 23-2, asked International President Al Hartung for an interpretation of the International constitution on Larsen's eligibility to hold office. On May 14 Hartung interpreted Article VII, Section 3, to mean that "even though a member has withdrawn from either the Communist, Fascist or Nazi party and is reinstated to membership in the IW of A, he still is not eligible to hold office in the IW of A."[35]

The case went to the first District 23 convention in May. Hartung records that Fadling and Larsen paid him an eleventh-hour visit in his convention hotel room "where (they) tried their best to convince me that as long as Larsen was no longer a member of the Party he should be allowed to remain an officer of the union. I would not tell them what I had ruled."[36]

After a bitter debate, the convention voted to disbar him, but Larsen appealed the president's ruling to the International Executive Board. His case was argued before the Board on June 3 and 4 by Walter Belka from Bellingham and Ray Glover from Enumclaw, Washington, but the Board upheld Hartung's ruling.[37]

Larsen then took his case to the 1953 International convention in Vancouver, British Columbia, but he was stopped at the Canadian border and prevented from entering the country to attend the convention. Belka, therefore, spoke on his behalf, pointing out that a 1945 amendment to the IWA constitution stipulated that a member must be tried before being expelled from office or from the union and that Larsen had not been tried by the IWA.

With the left wing still suffering the effects of the Canadian WIUC breakaway, there was little support for Larsen at the convention. B.C. delegates Geoffrey Amy and George McKnight championed Larsen's cause, as did an American, W.L. Harris from Reedsport, Oregon; their advocacy was to no avail. The convention voted to uphold Hartung's interpretation of the constitution and the decision of District 23's convention to vacate the vice-presidential position held by Larsen.[38]

The Aftermath of the Smith Act Trial

Five of the defendants in the Seattle Smith Act trial were convicted; one of them, Barbara Hartle, began cooperating with government authorities soon after being sent to prison. She identified scores of Northwest union and community leaders as Communist Party members. Some of the IWA members that she named were later subpoenaed by U.S. Representative Harold Velde's House Committee on Un-American Activities and interrogated at televised hearings in Portland and Seattle.

Radical unionists nobly stood their ground before the Committee and refused to be intimidated. IWA member "Brick" Mohr filled two pages of court record by refusing to answer a question about his place of residence. "You have my address. You served the subpoena," he told the committee at one point. When asked about his organizational affiliations, Mohr proudly told the Committee he belonged to "that organization that is on record with 60,000 members against the McCarthy committee, the Velde committee, and the Jenner committee...the International Woodworkers of America."[39]

The aftermath of the Smith Act trial and the HUAC hearings was devastating to the IWA and cast a pall of fear and suspicion over Pacific Northwest woodworkers. Communities which had been well known for their support of progressive politics now rejected all left-wing unionists, even those who were recognized for their established records of fighting for the interests of the working class. Karly Larsen, for example, found it impossible to rent a hall for his defence committee meetings in Stanwood, Washington, the town where he had worked for so many years. Larsen's son-in-law was fired from his job at a local dairy.

In the anguish of its defeat, the Communist Party turned inward and nearly rent itself apart. The Washington State CP Executive Board charged that Larsen had not conducted his defence "... on the basis of a principled fight against the frameup, but on the basis of retreat and capitulation." They criticized his District 23 convention position, saying that he had been poorly prepared and had failed to take his case to the workers to garner their support. Other defendants were criticized for "giving testimony about leaders and others not on trial" and for taking a "legalistic" approach to their defense. Party leaders readily admitted that the trial had taken a tremendous toll on the organization and its members, and that it had left the Party materially and emotionally drained.[40]

This was the climate in June of 1954 when 135,000 IWA and AFL lumber and sawmill workers cooperated in the largest shutdown in the Northwest's history. But numbers proved no match for the fear and intimidation that had pervaded union ranks. Despite help from unions across the country and a favorable ruling from the NLRB, the strike was lost. It would be over twenty-five years before woodworkers in the Pacific Northwest would challenge the industry again.

Restructuring Moves Forward

By the early fifties, the wood products industry was centralizing, and the International Executive realized that to keep pace with the growing number of corporate mergers it must move quickly to a regionalized system of contract negotiations. The IWA, however, was beset with structural difficulties which made such negotiations extremely difficult. These involved finances, administration and politics. Some districts, for example, were so small that they were a financial drag on the International, and the two or three financially viable districts resented having to subsidize the others to ensure their survival. These small districts were also a headache to administer, and they remained a political problem because it was in them that communist influence was strongest. An incentive to correct these problems at this particular time was the series of merger talks with the Pulp and Paper Workers which had so far been hampered by the IWA's unbalanced structure.[41]

With Larsen out of the way and the minds of the rank and file frozen with fear from the witch hunts, the anti-communists within the International leadership moved toward a final restructuring of the International which would preclude a resurgency of radical unionism. By abolishing the small districts, particularly in the Pacific Northwest, they felt that the communist influenced locals could be successfully reined in by the CRDC.

This final restructuring was the brainchild of Ed Kenney, the International's research and education director. He had been a negotiator for management against the CIO's Optical Workers Union in 1949 when Phil Murray had approached him about going to work for labor. Murray felt that Kenney could "stabilize" the IWA and prevent a communist takeover. In 1952 Kenney was sent to work for the IWA.[42]

The 1953 convention, citing the district council set-up as "an International within an International," authorized the International's president to appoint a committee to analyze the union's structure. When this committee failed to produce an adequate report in time for the 1955 convention, Ed Kenney was put in charge of the project. However, objections were raised to one person's being given sole responsibility for the work and to the instruction that Kenney develop "recommendations covering the complete program."[43] On the convention floor, Fadling suggested that staff people of other international unions had taken over and that woodworkers did not want to turn the destiny of their union over to the research director. A motion to strike the power to recommend from the resolution produced a 161-to-161 tie; it was then referred back to committee where this was struck. Resolution No. 26 (see Appendix II) then passed and subsequently became one of the most famous of all IWA resolutions.[44]

Kenney's earliest opposition came from the Northern Washington and Coos Bay Districts. They saw a pattern developing that was similar to that in the United Mine Workers, where District 12 was so large that, by controlling it, John L. Lewis was able to control the whole union. The Left feared that any regional council in the Pacific Northwest could be dominated by the CRDC. Under the present system, the districts had some autonomy and were able to use contract negotiations as a tool to build the union.[45]

Kenney's work also produced divisions in the conservative ranks where rivalry for leadership was rampant. It has, in fact, sometimes been suggested that anti-communism in the IWA was rooted from the beginning more in jealousy of Pritchett and other Communist leaders than in principled politics.[46] In any case, by the mid-1950s the anti-communists were fighting — even red-baiting — each other. (See Appendix III)

Kenney's plan put the entire Pacific Northwest under the control of the CRDC and strengthened the position of the latter in the International. British Columbia and eastern Canadian leaders immediately saw their own interests being frustrated by a CRDC monopoly of the International. Consequently, at the 1955 convention Harvey Ladd, from eastern Canada, attempted to scuttle the regional council proposal by moving to abolish the middle tier altogether and set up a "two-level structure composed of Local Unions directly affiliated to the International Union."

Although communism was still an issue, the line separating the Canadian social democrats from the more right-wing leadership in the Columbia

River District Council became increasingly important in the restructuring disputes. As late as 1954 there were rumors of another breakaway by the British Columbia District.[47]

But political lines within the conservative group were often blurred by the maneuvering to protect individual careers. Jim Fadling stood to lose the political leverage of a district office in the Kenney reorganization plan; he therefore moved that Districts 9, 11, and 23 be collapsed into one district. This would have given him and other western Washington leaders increased influence in the International; this motion was tabled.[48]

In 1957 Kenney made his preliminary report, and despite having been specifically instructed by the convention not to do so, he recommended changes. The IWA was weak, he said, because it was too democratic. In addition to the abolition of districts and establishment of five regional councils, he recommended (1) that International officers be nominated in convention by a nominating committee and elected in convention rather than by referendum, and that terms of office be changed from two years to four years; (2) that job stewards and plant committeemen be appointed rather than elected; (3) that business agents be appointed by regional executive boards rather than elected; and (4) that conventions be abolished at the regional level. The report also recommended substituting better collective bargaining techniques for strikes as a way of dealing with employers, and recommended concentrating upon large mills for organizing.[49]

The reaction to Kenney's report was immediate. Written critiques from British Columbia and the South attacked the undemocratic character of the report. Appointment of local and regional officials, and election of officers at convention, was called

> . . . something you would expect to see happening in pre-war Germany or Italy. To suggest that many local Unions need the service of a Business Manager appointed by the Region again gives credence and evidence to the trend that the report attempts to establish that of *appointment* over election.

> The day of free elections would be over. This smacks of craft unionism in more ways than one, for it is in the AF of L that you find Business Managers, not the CIO. It would seem that this type of dictatorial control is one of the points brought out at the McClellan Senate Investigating hearings in Washington, D.C.[50]

Abolition of district councils was called "the beginning of the end . . . democracy on an area basis is finished." The report left a general feeling that "what is being really attempted here is not so much to revamp the International, as it is to set up powerful regions giving the real control of the Union to the Regions."[51]

An avalanche of resolutions concerning Kenney's proposals hit the floor of the 1957 convention, several of them specifically addressing the issues of democracy in the union and political balance in the International. The strongest protests came from British Columbia, the South and District 23's President Jim Fadling. Lloyd Whalen, from British Columbia, expressed fear that if Kenney's report was adopted, "you wouldn't need any special committees or anything like that, all you need is a couple of carpenters to build a coffin, a few pallbearers, and that would be the end of the IWA."[52] Whalen's local moved, as an alternative to the five-region structure, that "there be two District Councils in the western states to cover the following areas: one District would cover all locals in Washington, Idaho and Montana and the other District cover Oregon and California." This would, of course, have broken up any CRDC control of the entire northwest. Fadling stated he was "not prepared to go ahead with the thought that the District Councils shall be abolished and Regions set up." It was a move, he said, "to start revamping this organization by duress, and not voluntarily."[53]

The convention's constitution committee combined all resolutions into one which called for a regional council set-up, the establishment of a committee to draft a new constitution, and a special convention to be held April 1, 1958 for the purpose of considering the new constitution. The committee was not united on the resolution, as was indicated by Alabama delegate Otis Matthews' attempt to submit a minority report. Matthews was prevented from doing so on a constitutional technicality and the convention adopted the committee's recommendation, thereby taking the restructuring out of Kenney's hands.[54]

The A-1 Committee (named for the resolution which established it) returned to the 1958 special convention with a proposed constitution which rejected many of Kenney's suggestions. Election by referendum was retained, as was the right of locals to choose their own officials rather than having them appointed.

The special convention opened with a fight over the procedure by which the rank and file would ratify the work of the convention. And there were accusations that district officers in the Pacific Northwest were attempting to stall the membership ratification process so as to postpone the decision until the next annual convention in 1959.[55]

However, debate over a change in roll-call voting by the International Executive Board produced most of the political fighting of the convention. At the time, each executive council member voted the strength of his membership on roll calls, and thus, the largest districts had the most power. The proposed change called for a simple one vote per executive board member. This debate highlighted the importance of structure in shaping the IWA. Joe Morris (District 1 president), who supported the change, argued:

> There is no place in the Executive Board for the use or misuse of power. We want no part of it. We believe that the smallest Regional Council in the International Union has every bit as much right to exercise their voice and has every bit as much right to be heard as the big Councils have. . . we believe this is the very essence of democratic decision [making].[56]

In a lengthy statement, Jim Fadling confirmed a great deal about White Bloc strategy through the years.

> How did we take this organization away from the Communist Party? When did we take it away from the Communist Party? *When we established the roll call vote*, when we eliminated the power plays by Harold Pritchett and his kind. Where would you people be at the present time if it wasn't for that? Yes, of course, I realize you can say that is in the past. When I went in the Navy, I thought it was in the past, too, but when I came back out of the Navy, less than two years later, I found two good old left-wingers where? As officers of this International Union. Yes, we gave them the old heave-ho. And how did we do it? *We did it by establishing District No. 13 in California, and we did it by establishing District No. 4 in the South so the membership would have a majority in a roll call vote in the Executive Board. That is the kind of power plays we used to keep this International clean, and I am proud of it, and I am proud of it . . .*

> Now, you might say the Commie threat is over. Is it? Weyerhaeuser Timber Company, Crown Zellerbach, regularly hire the few left-wingers that we have left in our organization and readily let them move around in our organization as they like, and the record is there. There is an underground movement. In Aberdeen at the present time, the leaders of the Communist Party back in the hungry thirties are again calling mass meetings of the unemployed, are again starting out the way they got their hold on the hungry workers when they started out. There is no doubt about it. The releasing of those last Commie leaders who were charged by the government has given impetus to the Red element that is still among our ranks, fostered and encouraged by the employers for one purpose — to attempt to tear down our union. There is no question about it.

> I say, in all seriousness, the roll call vote is not going to hurt you. If it is, then Pritchett was right; if it is, then Orton was right. If it is, then we, the people who grasped this organization away from the Communist Party, were wrong; and by God, I will never agree with that. Thank you.[57] (Emphasis added.)

Fadling's speech was to no avail. By a sizeable majority the convention accepted the change to one vote per council member on International Executive Council roll calls. Fadling at that point moved to reconsider a decision made earlier in the convention that representation on the Executive

Council should be on the basis of two members from each region. He moved that there be proportional representation; again it was to no avail.

There were several smaller debates over local rank-and-file control as opposed to centralized power. There was, for example, opposition to the requirement that appointment of research and education directors be approved by the International President, and there was opposition to the establishment of a pension plan for International, regional, and local officers.

The major restructuring — the formation of the five regional councils to replace the districts — was accomplished with little fanfare. In the interval between 1955 and 1958, peace seems to have been made among the contending rivals for International power, although at the present time, there is no record available as to when that agreement occurred and who supported and opposed it.

The formation of regional councils made the greatest difference in the northwest. District Councils 5 (CRDC), 6 (Klamath Basin), 7 (Coos Bay area), 9 (Plywood, Box Shook and Door), 10 (Eastern Washington and Idaho), 11 (Boommen and Rafters), 13 (Northern California), and 23 (Western Washington), and locals in Wyoming, Colorado, and Montana, were collapsed into Regional Council III, headquartered in Gladstone, Oregon (outside Portland). In the south, District 4 (Mississippi, Louisiana, East Texas, and Arkansas) was combined with locals in the southeastern states (Georgia, Virginia, and North Carolina) to form Region V. In the new setup, Region III (Pacific Northwest) became the largest with 35,000 members. Region I (British Columbia) had 30,000 members, followed by Region V (the southern states) with 17,000, Region II (Eastern Canada) with 7,000, and Region IV (the Midwest) with approximately 4,000 members.

The impact of restructuring varied from region to region. Little if anything changed within Region I, whereas in Region III, where one hundred and ten small locals had been amalgamated into one regional office, power was now centralized and moved further away from the rank and file. In the past the middle tier in the IWA structure had been a district office, with geographical, economic, and perhaps cultural identification with the ranks, but the regional council, which was to be the middle tier after 1958, was further removed in all respects.

The Kenney report had all the earmarks of classic business-unionism philosophy: control of all levels of the union through appointments by the International office; an organizing program that put dollars ahead of principles; substitution of collective bargaining techniques for class struggle at the point of production, and a shifting of the locus of power away from the local level. Kenney, by his own admission, had gone beyond the original intent of the 1955 resolution.[58] On the surface, therefore, it would appear that he was an isolated individual carrying out his personal designs for the reorganization of an International union. There are, in fact, still those who would maintain that Kenney's study had been intended as a springboard for his rise to the International presidency, but Kenney denies this.

It is more fruitful, however, to examine Kenney's influence in the context of the times, because the political thrust of his recommendations was consistent with the trends in the AFL and CIO during that period. The CIO was attempting to purge itself of its radical image in order to placate the witch-hunters and remain consistent with its anti-communist role in the international arena. In addition, CIO president Phil Murray was coming increasingly under the influence of the Catholic Church and following its more conservative and anti-communist political dictates. In the period immediately before and directly after the 1955 AFL-CIO merger, the CIO was forced to bring itself into line with the AFL way of doing things, and thereafter, the AFL's influence on internal union structures trickled down to the IWA through men like Ed Kenney.[59] Reforms that put voting strength on the basis of membership strength and replaced elections with appointments inherently propelled the organization toward conservatism.

Kenney's support came from within the Columbia River District Council, although even there it failed to unite into an effective force. Perhaps the internal rivalry for power which had been intensified by the abolition of several district-level posts precluded unity. It is significant, however, that this district which had been initially so hesitant about affiliation with the CIO, and which from the beginning had been the source of the strongest anti-communist sentiment, was also the district which supported Kenney's business unionism.

In a sense, the restructuring climaxed a period which was counter-revolutionary in the same sense that the period 1930–37 had been revolutionary. In the earlier period, the Communist activists who had been an organic part of the industry's mill towns and logging camps had led a movement which broke with the past practice of AFL business unionism and established a new union based on different principles of organization and leadership with different objectives. Almost immediately after the IWA's formation, however, the reaction had begun. Initially, the energy flowed from local pockets of conservatism in the Pacific Northwest and was given form and direction by individuals like Al Hartung, who were more pretenders for leadership than they were ideologues. But local currents converged with events on national and international levels. Through a series of interventions by the United States Government, the CIO, the Canadian Congress of Labour, and key activists like Adolph Germer, Bill Mahoney and Ed Kenney, the *Ancien Régime* had been restored.

Two decades of red-baiting and witch-hunting hung like an ideological albatross over the minds of the IWA's members. The fear of being labeled radical kept the membership from challenging conservative policies and blunted point-of-production militancy so that by the end of the 1950s, labor radicalism was dead in the wood products industry. It was succeeded by the same kind of business unionism from which radical woodworkers had sought to free themselves by joining the CIO.

7

The Legacy: Newfoundland and Laurel

Abandoned when the fledgling forest products industry moved its capital and labor west in the 1800s, the Eastern Seaboard and the South had been suffering from economic underdevelopment while the forests of the U.S. Pacific Northwest and British Columbia were being ravaged by clear cutting. And while class relations were maturing in the western states and provinces, the East and the South had remained non-union and politically unorganized. By the late 1950s, however, the relative attractiveness of the West had diminished. Unionism, albeit not the class-conscious unionism envisaged by the IWA's organizers, had become a fact of life in Oregon, Washington, and British Columbia; private forest lands had been cut over and the timber that was left on public lands was expensive to acquire. The East and South began to loom as important factors in calculating the industry's future.

The IWA's capacity to carry the struggle to new territory was conditioned by its experience in the West. The effect of the White Bloc's consolidation within the IWA could be measured in statistics by the end of the 1950s. In the Pacific Northwest, membership had dropped steadily after 1948, and in the four years following the disastrous 1954 region-wide strike, a thousand members per year had been lost. In Western Canada the union had only gained members by extending into Alberta, but it had nevertheless lost ground in the fight with employers. An eleven-union comparison of wage increases showed the B.C. District sliding from third place in the 1945–48 period to eighth in the 1948-51 period and tenth in the 1951-57 period.[1]

More importantly, years of factional strife had ended with the purge of its best strategists and hardest workers. In the absence of that leadership, it would be difficult to realize the potential for the revival of progressive industrial unionism inherent in the East and South.

Newfoundland

Newfoundland was a classic victim of uneven economic development. Until the twentieth century it had been captive to a one-industry economy — fishing; until 1949 it had been a colony of Great Britain. There was little domestic capital accumulation, and workers were dependent upon seasonal labor and frequent stints as migrant workers in other parts of Canada. In 1905 two British pulp and paper companies were attracted to Newfoundland by the huge tract of harvestable pulpwood and the availability of workers who could seasonally alternate their employment with the fishing industry.

Following Newfoundland's confederation with Canada in 1949, Premier Joseph Smallwood embarked on a campaign to "overcome the heritage of colonialism and merchant capitalism with a massive modernization and industrialization scheme."[2] Part of Smallwood's strategy entailed establishing an investment climate for resource extraction industries such as forest products, and by the end of the 1950s, the colonial bonds had begun to loosen as more diversified investments created new jobs for the emerging middle class. When combined with welfare expenditures, this helped to encourage workers to believe that they had a right to a better way of life. This was the state of Newfoundland's affairs when the IWA entered the province in 1956, perhaps expecting to ride this wave of economic expansion, while assuming that Newfoundland's lumber workers now dared to unionize.

The province, however, was a virtual captive of two pulp and paper mill giants: Anglo-Newfoundland Development Company (AND) and Newfoundland Pulp and Paper Mills Limited.[3] AND controlled 7,456 square miles of timber and maintained its own power plants and two railway lines; its corporate headquarters was situated beside the mill in Grand Falls, the largest of six company townsites. Grand Falls was a classic company town: AND owned 441 houses, the general stores, stadiums and hospital; there was no town council as the company governed directly. Corner Brook, in western Newfoundland, was headquarters for Sir Eric Vansittart Bowater's Newfoundland Pulp and Paper Mills (Bowater's), the largest manufacturer of paper and paper products in Europe and North America. The company owned the township, 800 houses and the "gentiles only" country club.

For twenty years, four unions — the Central Workers' Protective Union, Fisherman's Protective Union, Newfoundland Labourers' Union (NLU), and the Newfoundland Lumbermen's Association (NLA) — had been attempting to organize lumber workers on the island. Only the NLU and NLA had experienced a degree of success, but like the others, they were run as personal fiefdoms for the self-aggrandizement of their union presidents, respectively, J.J. "Joe" Thompson and Pearce Fudge.

In October 1956, the sixteenth convention of the Newfoundland Lumbermen's Association was addressed by Harvey Ladd, at that time still the IWA Director for Eastern Canada, and Andy Cooper from the United Brotherhood of Carpenters and Joiners. Both men had been invited by the NLA's Joe Thompson who had realized that with the merger of the CCL-CIO and the TLC-AFL to form the Canadian Labour Congress, the NLA would have to affiliate with either the IWA or UBCJ or be subject to raiding. When Ladd followed Cooper to the speaker's rostrum, Thompson attempted to influence the convention's reaction by introducing him as "the feller from the mainland."[4] Ladd lashed out against the Carpenters and proposed that a Newfoundland IWA local be established with loggers in full control. Later, delegates voted 27-16 for an *ad hoc* recommendation to affiliate with the IWA, but before the meeting adjourned, Thompson managed to get a 22-21 vote to delay the move.[5]

It was not an auspicious beginning for a campaign that could change the IWA's destiny; just 43 representatives had spoken for 14,000 Newfoundland loggers and several thousand more mill workers, and even then had only managed a split vote for affiliation. Nevertheless, the IWA Eastern Canadian District Council decided to go ahead with an organizing drive in Newfoundland. By December they had established a headquarters in Grand Falls.

With considerable dedication and tactical ingenuity organizers entered the woods in December 1957. With jeeps, four-wheel drive station wagons and snowmobiles, they headed up the rutted roads to the logging camps; they hired planes to drop leaflets into isolated sites and schooners to ferry organizers between fishing outports where many of the loggers lived. A helicopter and a ski-plane took organizers behind company boundaries after roads were closed. They supplemented direct contact with radio addresses that extolled the benefits that IWA loggers enjoyed elsewhere. Portable generators were transported to remote camps to screen pro-union films. When not in the bush, staff members helped loggers file unemployment insurance claims.

Newfoundland offered the IWA an opportunity to redeem its dismal record among loggers in Quebec and Ontario during the late 1940s. Harvey Ladd knew this, and he also understood that success on the island could open the door to the IWA presidency.[6]

By March 1957, eighty-seven percent of AND's woods workforce had signed petitions requesting a representation election, and the IWA filed an application before the Newfoundland Labour Relations Board. The Board brushed aside the application because the IWA had not chartered any locals in Newfoundland at the time of application. In June the union returned to the Board with petitions from ninety-two percent of the loggers and two locals chartered. The Labour Relations Board ordered an election between the IWA and the NLA. Balloting by mail in the widely-scattered camps and outports lasted from January 30 until April 12, 1958; when the votes were counted, AND loggers had chosen the IWA 5,197 to 498.[7] Seeing the inevitable end, Thompson took the remnants of his NLA and its assets into the United Brotherhood of Carpenters and Joiners in June and set up UBCJ Local 2654 in Grand Falls.

The 1959 Strike

Delegates from thirty-eight AND logging camps met in Grand Falls, May 29 to June 4, 1958 to draw up a wage and working conditions policy; they presented it to the employers on June 17, the day that negotiations opened at the company offices. By October they had reached a stalemate, and the union asked the government to appoint a conciliation board. Two months later the board made major recommendations which included a five-cent-an-hour increase over a two-year contract to $1.22 per hour, a second-year work week reduction from sixty to fifty-four hours without loss of pay, and a union shop. Safety and camp living conditions were to be resolved cooperatively in discussions between the union and management.[8]

The Newfoundland Minister of Labour urged acceptance. The AND Company's representative on the three-man board signed the report making it unanimous.[9] The IWA accepted the Board's recommendations and was "...willing at any time to negotiate an honorable settlement so that both the men and the company [could] prosper..."[10] But AND remained firmly opposed. Ninety-eight percent of the 1,300 loggers actually at work in the company's camps voted to strike at midnight, January 1, 1959.[11]

The smashing victory in the April representation election had undoubtedly given Ladd the confidence he needed to go ahead with the strike. But although the loggers were clearly dissatisfied with their working conditions and tired of the NLA, the IWA was not prepared for a strike in Newfoundland. By failing to form locals at the very beginning and then assisting those locals to organize the election of their own representatives, Ladd had deprived the loggers of a crucial learning experience and an opportunity to tighten the bond between the union and the community. The two locals eventually formed had only 500 members, or less than ten percent of the loggers, on the eve of the strike.

Ladd had collected signatures and turned people out for an election, but getting them to vote and keeping them on the picket lines in the middle of winter were two different things. During the two years of Ladd's campaigns, the International had given virtually no coverage to it in the International paper. There was no strike fund. Ladd and three others on the International payroll had done nearly all the organizing, and they were constantly baited as outsiders by the company and other unions. With Newfoundland workers having played no role in setting the plans that now propelled them toward a walkout, and with only ten percent of the island's loggers and none of its mill workers having voted for it, the base of support was clearly too small to win one of the greatest strikes in Canadian history.

A brief period of quiet and orderly picketing followed the establishment of the IWA's lines around logging camps and along forest roads. Many of the loggers were counseled to simply remain in camp as the strike might be quickly settled, but the company, fearing property destruction, charged them with trespass and kicked them out, although not before they had eaten one last meal at company expense. Most loggers manned the lengthening picket lines, but a few returned home.[12]

Pickets soon stretched for two hundred miles along the central Newfoundland section of the Trans-Canada Highway and along the lesser roads that connected Grand Falls to the four woods divisions of AND at Terra Nova, Bishop's Falls, Badger and Millertown. AND countered by hiring fishermen from distant outports and high school students as scabs. Loggers watched as car convoys with R.C.M.P. escorts pulled off the main highway and headed past them into the woods on the company's private roads. The *Grand Falls Advertiser* began a campaign against the union by editorializing:

> Do you honestly think these leaders in their swanky hotels are interested in the Newfoundland logger and what he gets and how he lives? No, unfortunately there are too many James Hoffas, too many Dave Becks in labour circles nowadays.[13]

The editorials in the twice-weekly paper were so unabashedly anti-IWA that AND reproduced and circulated them.

Ross Moore, president and general manager of AND, plied the business community with hints and threats that if the IWA succeeded, AND would shut down and relocate, leaving Grand Falls and Newfoundland to wither. The AND mill began using the mini-mountains of pulpwood stored earlier in the company yards to wait out the strike. The mill workers, who had never been part of the strike strategy, kept working while striking loggers were soon dependent on the resources of the union. And while the union had pledged that the strikers would not go hungry, sufficient money had not been raised to feed them. It was not until April, when the strike was nearly lost, that the Canadian Labour Congress announced plans to raise a million-dollar strike fund. The largest appeal of its kind in history, it eventually raised $856,000, with over half of that coming from Canada, but as a morale booster, it was too late. Cold and hunger had set in long before. Cutting gangs had been organized to provide free firewood to strikers' families, and the union had issued vouchers – $5.00 per week per man, plus $5.00 per wife and $2.00 per child – to be exchanged at authorized food stores, markets and clinics. But the weekly strike expenses of $48,000 for vouchers, legal fees, transportation, picket line maintenance and staff expenses used up donations as fast as they came in. [14]

Enter Joseph R. Smallwood

The strike had been in effect for over a month with no sign that the union was giving up, when Newfoundland's Premier Joe Smallwood intervened to end it, fearful, perhaps, that it would extend into the spring cutting period. A former labor organizer, Smallwood had been premier since 1949 when he had led the movement for confederation. [15] He was the island's lynchpin, its political champion, the new king of the old colony. Adroit and manipulative, he was the arbiter of Newfoundland's immediate past, and he claimed most of the credit for the benefits confederation had brought to Newfoundland, credit which should have gone to the federal government. He so dominated radio (and television when it came to the island in 1959) that his political fiefdom had been unchallenged until the IWA moved into Newfoundland.

Surprisingly, Smallwood had maintained his silence during the initial weeks of the strike, but with the strike lingering, he could wait no longer. He declared himself Minister of Labour and then, on the evening of February 12, tore into the IWA in a province-wide television and radio address. [16]

It was a masterful speech. Smallwood clothed himself in pro-union rhetoric, paternalistically evincing his concern for the loggers' welfare and advocating a strong union for the woodworkers. But, he claimed, the confidential reports from clergymen, concerned citizens and people he had sent to the outports and settlements to be his "eyes and ears" during his period of silence had convinced him that the "ordinary" people wanted him to intervene against the IWA. As a good premier he was simply following the will of Newfoundlanders. Of course, he said, if the choice was the IWA or no union, then even Smallwood, if he were a logger, would ". . . be just as strong for the IWA as any other logger." But the premier magnanimously offered his help to form such an alternative if the loggers would only follow his advice and ". . . send the IWA out of Newfoundland."

Smallwood thus conceived the Newfoundland Brotherhood of Wood Workers (NBWW), and he picked Max Lane, general secretary of the Fishermen's Federation and a member of the House of Assembly, to lead it. The full weight of Smallwood's government would be placed behind the NBWW. If the loggers would support the new union, they would be back to work in one week and have an agreement with AND within a fortnight. Clearly the government had replaced AND as the union's chief adversary.

The Premier concluded his speech with eloquent demagoguery:

> The government don't want the IWA. The government will never work with the IWA and will never talk to them, will never answer a letter or telegram from them, will never have anything to do with the IWA . . .
>
> How dare these outsiders come into this decent, Christian province and by such desperate, such terrible methods try to seize control of our province's main industry. How dare they come in here and spread their black poison of class hatred and bitter bigoted prejudice. How dare they come into this province amongst decent God-fearing people and let loose the dirt and filth and poison of the last four weeks. The very presence of the IWA in Newfoundland tonight is an insult to every decent Newfoundlander.[17]

On February 26, Smallwood christened the NBWW at the Grand Falls Town Hall, but his new union held little attraction for the loggers. The kickoff attracted fewer than the expected 700 people; most of those who attended were local pulp and paper mill workers and government employees who may have come for the free coffee, sandwiches and doughnuts, or for the reimbursed travel costs if they signed an NBWW application card. That same afternoon, however, 1,200 striking loggers turned out for an IWA counter-meeting in Bishop Falls.[18]

The next day eight major unions of the Grand Falls-Gander District Labour Council, representing nearly 3,000 members, refused to comply with a CLC request for full support of the IWA; the Congress immediately dechartered the Council. This was another indication that the Grand Falls–Gander pulp and paper workers did not support the loggers in the IWA.

The IWA Outlawed

Smallwood returned to St. John's from Grand Falls to reconvene the legislature, and for three days, beginning March 4, he dominated the sessions with a demonic harangue against the IWA.[19] The purpose of the special session was to pass two bills which would force the IWA, the certified bargaining agent for AND, out of Newfoundland and establish the NBWW as the only "union" open to the loggers.[20] Bill 1, titled "To Make Provision for Safeguarding the Public Interest in View of the Present Unsettled Conditions in the Woods Labour Industry," would decertify IWA Locals 2-254, Grand Falls, and 2-255, Corner Brook. Without certification, the IWA could not force the companies to negotiate. Smallwood's Bill 2 associated the IWA with the alleged crimes of Jimmy Hoffa's Teamsters in the United States, then under investigation by Special Counsel Robert F. Kennedy and the McClellan Committee in sessions which were widely covered by television. Accordingly, the Newfoundland Labour Relations Act was amended to read: "[a Newfoundland union], a substantial number of whose superior officers have been convicted of such heinous crimes as white slavery, dope-peddling, manslaughter, embezzlement, such notorious crimes as these, [will be decertified by the government.]" The ease with which the two bills passed is a credit to Smallwood's total control of the legislature.[21]

Newspaper coverage soon reflected the legislative proceedings. Ed Finn, the dynamic, twenty-eight-year-old editor and columnist of the Corner Brook *Western Star*, had been running columns, editorials and cartoons favorable to the union;[22] after the special session, all his columns dealing with the strike were reviewed before publication. But when the newspaper's head office in St. John's vetoed an IWA advertisement, Finn resigned. His staff followed suit, and Newfoundland was left with papers which dutifully printed the company's press releases and either ignored, or editorialized against, the IWA's position. Smallwood could therefore easily brag that ". . . every newspaper in Newfoundland. . . supported [me]."[23]

Death in Badger

The strike's first violence occurred in Badger, a loggers' town and a center of power for the IWA. As in other towns, the union had established three to four billets there,[24] as central housing was a necessity for effective picketing when the strikers had to gather from such widely-scattered towns and outports. One large house in Badger was packed with loggers who had left Bowater's woods in the west and moved to Badger to aid the union. From their billets, the loggers could fan out to stop scabs from coming down the highway or to block the forest roads that branched south from Badger towards Millertown and Buchans, a principal cutting area. If AND, with help from the new NBWW, could get men past the Badger picket lines and back into the woods in early March, it would both break the strike and meet the mill's demand for pulpwood.

On March 10, the company attempted to move strikebreakers to Miller-town. Smallwood authorized a squad of St. John's police to augment the R.C.M.P. headquartered in Grand Falls, and by 6:00 pm a force of forty-five Mounties and fifteen constabulary had made the trip from Grand Falls to Badger. R.C.M.P. Inspector Arthur Argent, commander of the contingent, arranged the men three abreast and twenty deep, and marched them west up Badger Road in the quickening dark. Ahead, some two hundred loggers, their wives and children and onlookers blocked the Buchans-Millertown intersection. They were ordered to clear the road.

The swearing and defiant shouts of the loggers swelled as the formation approached, but the crowd opened up, and the Mounties and constables marched through. Then the column reversed, the police marched back down Badger Road, and with a sudden sharp, column right-turn moved into Buchans Road where the loggers had bunched. Police and loggers were unexpectedly face to face with little room or time for the loggers to fall back. The police broke ranks and struck first. The loggers wielded birch clubs snatched up from nearby stacks of firewood or sticks they had hidden inside their pantlegs. Loggers in dark parkas and green tartans mixed with police in blue uniforms and grey longcoats.

Some of the loggers jumped the snowbanks and retreated down alleyways pursued by police; a few stood and slugged it out, warding off the batons and crops.

Logger Ronald Laing's son dropped his wooden club and ran across Badger Road into the alley beside the Full Pentacostal Church where he was cornered and knocked to the ground by the police. Laing jumped to his son's rescue but as he moved to pull the police off the boy, he was brought down, unconscious. A moment later, a birch club slashed through the air and caught St. John's constabulary officer William Moss solidly enough to kill him.

The battle was over in less than fifteen minutes. Nine loggers were arrested, but the person who either threw the birchwood or struck the blow that killed Moss, was never identified. Within twenty-four hours Premier Smallwood pronounced his own verdict: ". . . the real criminals were not arrested last night. All Newfoundland knows who the real criminals are, [IWA organizers] Ladd, Hall and McCool."

Moss' funeral was well-orchestrated. Three hundred Canadian Legionnaires marched in front of war veteran Moss' coffin from the AND's company hospital through Grand Falls to the CNR railway station. Shortly after the casket had been placed aboard the train, the crowd turned on the nearby IWA office. Bricks, bottles, and chunks of concrete crashed through the windows. The R.C.M.P. stood by, mingling with the mob, but arrested no one. In St. John's, the flag-draped casket, accompanied by Smallwood and almost the entire Newfoundland government, moved slowly and silently, along the crowd-lined streets from Moss' home to the railroad station. With the body aboard, the train stopped or reduced speed at all the villages on the return trip for burial at Port Blandford, Moss' birthplace. Smallwood had effectively turned the burial into a two-day media trial of the IWA, almost placing the union into the grave with Moss.[25]

The Newfoundland Brotherhood of Wood Workers and the UBCJ

While the IWA was struggling with the effects of Moss' death on its membership, the Newfoundland Brotherhood of Wood Workers quietly conducted negotiations with AND;[26] on March 12 an agreement granted a first-year, five-cent hourly wage increase — exactly what AND had refused the IWA three months earlier. Bowater's followed suit. The companies would also now act as the bargaining agent for their sub-contractors, something that they had refused to do for the IWA because they had been trying to force the union to negotiate with each individual sub-contractor in order to stop or retard certification of the union. The new policy could end the sympathy walkout by Bowater's loggers and get crews back in the woods to complete cutting for the spring drive in the west.

It had been a long, cold winter of hunger and insecurity for the loggers and their families and the men were anxious to get back to the woods, but to get work, they had to join Smallwood's NBWW, clearly a company union. In fact, Article II, Paragraph 3, of the NBWW constitution permitted suspension of loggers who retained their IWA membership.[27] And instead of an NBWW representative at the job site, an AND camp foreman would enforce the mandatory NBWW membership. These regulations, when coupled with AND's lengthy blacklist and the government's withdrawal of welfare allotments to IWA members, gave the NBWW an enviable handicap.

In March, 1962, the NBWW was formally absorbed into the UBCJ, and working agreements were signed with the paper companies. Stirling Thomas, a mill worker from AND who had replaced Lane as NBWW president, returned to the company and received a promotion to plant security officer. The other seven officers of the NBWW shifted directly to the UBCJ.[28]

Perhaps there was no way Newfoundland could have been won for the IWA, given the province's colonial character, but the union had also made crucial mistakes. By moving too quickly and not building on local institutions and leaders, it had left itself open to the serious charge that it was led by "outsiders." Only one of the four organizers, Hank Skinner, was from Newfoundland, and he was the only one that the International had objected to.

The conventional wisdom says that the IWA had made the mistake of taking on the whole fabric of Newfoundland society. But in fact, that's exactly what they had not done. They had conducted a wages, hours and working conditions strike with a small base in one occupational sector. Their efforts did not approach the society-wide level of struggle called for in Newfoundland; they did not even represent good industrial unionism. Unfortunately, the realization that a more widespread political struggle was necessary came far too late. That kind of campaign would have needed the total support of the International to succeed, but the International's leaders were typically business unionists in their reluctance to invest in long-term effort.

In the end, the greatest asset the IWA had in Newfoundland — the charismatic leadership of Harvey Ladd — may have been its fatal weakness. The International's executive feared that Ladd was building an empire in the east to bolster his pursuit of the International presidency out west and was, therefore, reluctant to give him help. And he was too powerful for Smallwood to allow to succeed. Ladd's élan had been substituted for prior planning and patient organizing, and when the strike was over and he went back to the mainland, he left nothing behind. The Newfoundland situation had called for radical strategy and leadership, the kind that had organized the Pacific Northwest and Vancouver Island years earlier. But the radicals were gone, and Newfoundland was lost.

Laurel

Ten years after its Newfoundland defeat, the IWA found itself on the ropes again, fighting for its life in Laurel, Mississippi. Unlike Newfoundland where dictator Joey Smallwood's sledgehammer tactics clearly defined the opponents, the Laurel situation involved delicate racial politics, sophisticated legal maneuvering, and terrorism.

The wood products industry had overcut the South in the 1800s, but as the timber grew back and the workers grew hungrier, the industry again found the South an attractive investment scene. Post-World War II technology made it possible to manufacture plywood from small southern pine logs. As a consequence, between 1958 and 1976 the percentage of U.S. plywood coming from the northwest's giant Douglas fir trees fell from ninety-seven to fifty-three percent. Low wages prevailed in the southern industry: in 1971 the average hourly wage for lumber workers in Oregon and Washington was $4.31, compared to $2.36 for the largest lumber-producing southern states.

The Laurel local had been a keystone of the IWA's Southern program since the late 1940s. When the CIO had opened its "Operation Dixie" campaign to organize the South on March 3, 1946, the IWA had been able to plan on forming a Southern District Council for the first time in several years. During that summer the IWA unseated the AFL's Pulp and Sulphite Workers at the large Laurel Masonite plant, and within weeks had secured a union shop agreement.

Operation Dixie was short-lived, however. With the passage of the Taft-Hartley Act in 1947, organizing in the South became more difficult. Moreover, the division within the CIO which followed the implementation of Taft-Hartley's anti-communist provisions, diverted attention from organizing to purging the union of its leftist elements. The fight between the Red Bloc and White Bloc within the IWA which culminated in the secession of the B.C. District from the International had left southern woodworkers fending for themselves while the International poured thousands of dollars and dozens of organizers into a campaign against the communists in British Columbia.

Laurel became the symbol of a fiercely independent local battling the intrusion of northern monopoly capital. In 1949 the local struck Masonite for five months over wages, length of work week, grievance procedures, seniority, and vacations. With the slogan "No Contract — No Work," 2,600 IWA members walked out at midnight, April 1. In August Masonite tried to organize a back-to-work movement, but when this was unsuccessful, they had little choice but to settle with the union in early September.[29]

The struggles in Laurel were only beginning, however. In July 1952, 365 workers at the Mengel Lumber Company struck, but the company rallied community and business support to its side when it threatened to close for good and seek cheaper labor elsewhere.[30] In a letter on September 26, plant manager Robert Hess told employees that if they wanted a union, "they might give the Carpenters a trial."[31] The Carpenters then joined the attack on the IWA, claiming the action at Mengel's was not a strike but the work of one individual — J.D. Jolly, president of IWA Local 5-443. The Carpenters issued leaflets making personal attacks on Jolly,[32] but the Laurel rank and file stuck by their president, and the strike continued. Then on December 5, 1952, the IWA won a representation election over the AFL by an 8 to 1 margin. Four months later, the local was able to announce a victory with "one of the best contracts ever."[33]

Still, unionism in Laurel could not develop in a vacuum, free from the fundamental contradictions that were maturing in the United States in the mid-1960s. Labor and civil rights — two of the most powerful social forces in the twentieth century — were bringing a social revolution to the South.

Laurel in the 1960s

By the early 1960s, ɪᴡᴀ Local 5-554, with its nearly 3,000 members at Masonite's Laurel operation, had become one of the most militant union locals in the entire South. The company could match its resourcefulness at the bargaining table, but the workplace belonged to the union. The key to its power lay in a seniority system that prevented the company from eliminating or combining job assignments. Numerous times the local had called wildcat strikes over shop-floor issues, and each time it had won.

But competition within the increasingly monopolized wood products industry was forcing Masonite to modernize, automate and increase productivity. With the giants of the industry — Weyerhaeuser and Georgia Pacific — moving operations into the South, more capital-intensive processes and a better-controlled — even diminished — work force was essential if Masonite was to stay competitive. Management decided that the cocky local union would have to be dealt with once and for all.

Masonite was the largest employer in Jones County, where racial antagonisms had been seething for years. The county was a primary target of the civil rights movement, and Laurel, reputed to be the headquarters of the Mississippi Ku Klux Klan, was the bullseye. (Some observers claimed the Klan ran the ɪᴡᴀ Local.)

Beginning with Presidential Executive Orders in 1961, the Federal Government began pressuring employers to integrate their work forces and end discrimination in hiring and promotion. Masonite maintained separate locker and shower room facilities and segregated employment offices. The changes required by the government would threaten a system of management practices and dominant racial ideology that were decades old.

The working class community was also part of this same social system. A dual seniority system — the same seniority system that allowed the union to control workplace politics — favored white workers. The union local had an all-white leadership with a separate sub-local for black workers; when the union met in convention, black members had separate hotel accommodations. Thus, if compliance with civil rights law meant an end to Masonite's divide-and-rule racial policies, could it not also mean an end to white working class privilege, an end to the seniority system and an end to the union's control of the workplace?

It was obvious, at least to the company, that forced integration could be an explosive issue for the workers. Where the charges were placed and when the fuses were lit would determine who got hurt. Masonite struck the first match in March 1964 by eliminating dual seniority lists and integrating facilities. In May of 1966 it announced its intent to make more changes and began an automation plan that eliminated jobs by combining job classifications.[34] Meanwhile, a suit was filed against the union local forcing abolition of the black sub-local and the holding of new elections. The newly integrated local was seventy-five percent white; in the elections which followed, none of the black candidates were voted into office. The resentment of the black workers simmered while tensions building toward a strike against the company began to mount.[35]

Even more volatile pressures were building in Laurel. Early in 1964 Sam H. Bowers, a local 40-year-old Klansman known for his "rather strange habits — like wearing swastikas on his arm or clicking his heels in front of his old dog, stiffly throwing a Nazi salute and exclaiming, 'Heil Hitler,' — formed the White Knights." Within four years, Bowers had become a suspect in 300 acts of murder, beatings, bombings and burnings.[36]

At its peak Bowers' "hooded order" numbered 5,000 to 6,000 members, who conducted a systematic campaign of terror in the Laurel area, much of it being directed toward the civil rights activists from the north. The most infamous murders took place June 21, 1964 when three civil rights workers — Michael Schwerner and Andrew Goodman from New York, and James Cheney, a Meridian, Mississippi black — all in their 20s, were shot by White Knights in a conspiracy aided by law enforcement officers. Forty-four days later, the bodies of the three men were pulled from a dam near Laurel as the entire nation watched on television. Bowers was convicted of the killings but released while the decision was appealed.

In November of that same year, Otis Matthews, the white financial secretary of Local 5-443, was viciously attacked by the White Knights because they "believed he was promoting integration" at the Masonite plant. Eight Knights using a "black annie" — a wooden-handled leather strap four feet long, one-quarter inch thick and four inches wide with small holes in it — beat Matthews nearly to death.

Racial tensions, needless to say, were running high when a new contract was signed on March 3, 1967 between the union and Masonite. Tension was increased because Masonite's automation program and job reclassification schemes were beginning to show results, with a drop in employment of 1,000 workers in the previous year and a half. To make matters worse, the company began "Klan-baiting" the local, charging it was run by the Ku Klux Klan. Although Klansmen were members of the local, local officers had bought a full-page advertisement in the Laurel *Leader-Call* and "threatened war on the Klan" at the time of Otis Matthews' beating. J.D. Jolly, president of the local, warned the company that there would be a strike if it didn't halt its campaign against the union.[37]

The company played up to its black workers, pointing out its record in ending segregation practices. "When management was confident that it had the loyalty of enough black workers, it began upgrading black workers in a way that was certain to drive white workers into a wildcat." Then days before the union gave any indication it planned to strike, Masonite began hiring Wackenhut guards and setting up sandbag machine-gun nests. Clearly, Masonite was exploiting the strife between union members to set the stage for a strike it wanted — and would precipitate.[38]

On April 21 two workers were asked to do work not in their job descriptions. When they refused, their shop steward Freddie Adcock backed them and was promptly fired by the foreman. The strike was on. "That strike was set up and Jesus Christ himself couldn't have prevented it," declared Granville Sellers, a local vice-president.

Masonite telegraphed International President Al Hartung on April 23 requesting that he "take immediate and appropriate action to return all men to their job." Hartung ordered Local 5-443 to return to work. Confusion reigned. The company had violated the contract and the strike was legal.

Hartung's receptivity to the company's plea was the culmination of a tragedy thirty-five years in the making. Hartung, the one-time camp foreman, had wanted to be International President since the formative years of the union. He had been defeated by Harold Pritchett for the Presidency of the Federation of Woodworkers in 1936 and defeated by Pritchett for the IWA Presidency in each of the next two years. Hartung had found his political base outside the union in the CIO bureaucracy and with that leverage had been able to win the top post by a narrow margin in 1951. The Laurel situation offered Hartung his chance to prove his mettle, a chance to prove he was indeed as good an organizer and leader as the radicals he had fought for so many years. It was a situation that held the same promise for Hartung and the union movement of the late 1960s as the unemployed and relief movements had held for the radicals of the early 1930s. By merging the interests of the union with those of broadbased community concerns, the union could have established its relevancy for a new generation of workers and replenished its ranks with a fresh cohort of activists.

But Hartung and the Communists who built the union were not cut from the same cloth. Hartung's Laurel action was consistent with his middle-class background, his record as a business unionist and his reputation as a leader of the IWA's anti-communist bloc. Success for him was measured by the traditional standards of individual status and security; being a successful trade unionist only meant personal advancement. But he could advance no further than the office he already held; for him to have turned to a strategy of popular mobilization in a time of crisis would have been totally out of character. So Hartung embraced Masonite, presumably gambling that the company would reward his acquiescence with a cooperative attitude toward the union later on. As a result, the International stood in the company's corner, while the local continued the strike alone.

Hulse Hayes, a lawyer from the union-busting firm of Senator Robert Taft, went to court for Masonite to seek an injunction against the union. A lower court judge ruled in favor of the union but Hayes appealed to the Mississippi Supreme Court and got his injunction even though subpoenaed company records "showed the company had hired Wackenhut guards 10 days before the strike." If the company hadn't forced the strike, why had it hired armed guards?"[39]

Masonite dug in. The plant "looked more like a fortress than a factory." Scabs were recruited from other parts of Mississippi, Georgia and Texas. "They had the whole Ole Miss football team scabbing in there," declared Jolly.[40]

The company's race-dividing tactics began to pay dividends. One black union member said:

> You know, after the Civil Rights Act the company integrated the facilities at the plant. Washrooms and showers. The whites kicked up a fuss.
>
> I don't appreciate a man using me...saying we're together when there's a strike and then kicking up a fuss because we're using the same facilities. Another thing, the whites didn't even consult us when this strike began. Not one word, like they figured we'd support it anyway. I didn't desert the union; the union deserted me.[41]

Another black member complained of being treated "like a second-class citizen, like niggers."

> Ain't no Negro union officials. There's only four black shop stewards, and they don't have authority over whites.
>
> The union never asks us for help until something like this comes up. To put it right down, Negroes are just sick and tired of being fooled by whites. But no more. I'm so tired of my people being walked on.[42]

"Klan-baiting" was paying off. "It's a den of Kluxers," Mrs. Susie Ruffin, a veteran freedom fighter and leader of the Mississippi Freedom Democratic Party, said of Local 5-443. No matter that the local's leaders had taken a militant stand against the Klan only three years before. The Mississippi AFL-CIO President Claude Ramsay concurred that "Kluxers" had taken over the Laurel local.[43]

Black workers began returning to work intensifying the race hatred even more. Violence raged. The homes of black workers were regularly riddled by gunfiire.[44] Five men died before the strike ended.

One June 30 the NLRB dismissed Local 5-443's charges against the corporation, and on July 12 it dismissed Masonite's charges against the International, but at the same time issued a complaint against the local. The local was now pursuing an illegal strike but they had wildcatted before and never been beaten, so on July 5, the membership rejected the company's proposal of "super seniority for those who had already returned to work and union acceptance of all changes in work assignment."[45]

Meanwhile, another series of events was unfolding that would produce the strike's most important consequences. On June 26 Robert Paul, Masonite General Manager, had requested the International to "appoint a receivership for the Local." On July 17 the International's attorney, James Youngdahl, joined the call for the International's seizure of the local. "I think it is apparent," argued Youngdahl, "that the company is not going to give in and that the possibility that the local union being destroyed permanently and the International Union harmed greatly by [a] damage suit is very great."[46] From that point on the local's interests were pitted against the International's: would the strike be won by the local or settled on the company's terms in order to save the International? In October the International moved to settle the strike behind the backs of the militant Local 5-443 members.

The local had gained allies of its own by this time — allies which complicated the already complex politics of the strike. The struggle had been joined by GROW (Grass Roots Organizing Work), an affiliate of the Southern Conference Educational Fund (SCEF). GROW was staffed by white veterans of the civil rights movement, many of them former members of the Student Non-Violent Coordinating Committee (SNCC). GROW had been conceived by former SNCC field secretary Bob Zellner, his wife Dottie and civil rights activists Carl and Anne Braden, with black-white working class unity one of its organizational objectives. If it could be accomplished in Laurel, it could be accomplished any place, so GROW accepted the challenge.

They began holding meetings with the strikers with the immediate objectives of getting the scabbing black workers out of the plant, and encouraging white workers to allow the participation of black workers in union affairs.[47] The local's white leaders saw that with GROW's involvement their race-divided local could be pulled together for a fight against both Masonite and the International. A bi-racial committee of five was formed to lead a rank-and-file movement for control of the strike.[48] GROW's offer of legal assistance to fight the International's seizure of the local sealed the alliance between the civil rights activists and the supposedly Klan-infested local. In November the rest of the civil rights movement came over to the side of the beleaguered local. The Mississippi Freedom Democratic Party denounced Masonite's Klan-baiting tactics and called for black-white unity around Local 5-443.

> If Masonite can kill the union in Laurel, this will set back both Negro and white workers 30 years. . . .

> Masonite claims that the union is completely controlled by the KKK. This places Negro employees in the unfortunate position of having to choose between the KKK or Masonite. This is too simple and too frightening to be true.

> It is to Masonite's advantage to emphasize the KKK element in the union. This keeps black and white workers apart. . . . We refuse to swallow Masonite's claim of complete KKK control of the union. We have found that labels, whether communist or KKK, prevent people from thinking and cover up the real issue.

> The Klan issue should not force Negro workers to reject unionism.[49]

But unity came too late. By September over 1,000 workers, black and white, had returned to work. The International was negotiating directly with the company, which took the position "that it had won the strike [and] had no need for any of the strikers." Finally, negotiations were moved completely away from Laurel: on November 15 company lawyers and International lawyers met in Chicago where Masonite was headquartered. There the International agreed to give up its rights to challenge changed job assignments (the original reason the local had struck) in return for some seniority rights for strikers. It was nearly the exact settlement the rank and file had rejected in July. Over the objections of J.D. Jolly, the International Executive Board voted on November 28 to settle the strike, and on December 6 in San Francisco, the International and Masonite signed a contract for the Laurel workers.

The strike was over but the fight between Local 5-443 and the International was not. One of Masonite's conditions of settlement was that the local be placed under trusteeship. The International had agreed, so the day after the contract was signed, the local's charter was withdrawn.[50] J.D. Jolly was removed from office, and all property and finances of the local came under the control of the International.

Hoping to sooth the pain, if not co-opt the anger of the deposed local leaders, the International offered Jolly the trusteeship of the local. Jolly, who had helped organize the local and led it for over twenty years, denounced the offer as hypocrisy. He attacked International President Ron

Roley as "a politician without honor" and called the settlement "dictatorial."[51] Rank-and-file members called the final disposition "a sellout, a sweetheart contract."[52] The failure of the strike left blacks and whites divided even more. "Our local leaders might have failed," wrote Ray Craft, "but your International not only failed us, they broke us and scabbed us. So I can promise you one thing, if I have anything to do with it, we will reorganize, but in time we will get out of your scabbie organization. Any other man or woman that is affiliated with your organization had better get out before you sell them out too."[53]

With the help of GROW, a group of local members went to court to overturn the trusteeship. The petitioners argued that the International had violated the due process provisions of its own constitution by seizing the local. The court disagreed and the local stayed in trusteeship until September 17, 1969 when newly elected officers resumed control.[54]

Laurel and the 1970s

Technically, all matters concerning the Laurel strike were closed by 1970 but the aftershocks of the strike reverberated for a long time. The settlement and trusteeship left many members and former members of the local bitter. A "Committee for Better Union Leadership" was formed to carry on the struggle for justice in Laurel and agitate for reform of the IWA. The committee included Granville Sellers, an ousted vice-president of Local 5-443, and several hundred black and white workers. With the help of SCEF, the committee petitioned the U.S. Justice Department to investigate a possible conspiracy between the IWA and Masonite.

The unanswered questions of Laurel followed the IWA into the 1970s, perhaps the most important question being, What did Laurel mean for the IWA's future? The challenges of Laurel in 1967 had been an opportunity for the IWA. The union could have joined hands with the radical civil rights movement and boldly stepped into the future, because that movement, like the unemployed movements of the early 1930s, was a training ground out of which a new generation of organizers could come. But for the union's leaders, the civil rights movement bore far too much similarity to the left-wing movements they had fought for so many years.

The IWA may also have turned a corner at Laurel. Mistakes or not, hero or scoundrel, the IWA justified what it did in Laurel by professing to have fought reactionaries, not progressives; it had broken a den of Kluxers, not a cell of "commies." This willingness to project a public image as a force for progressive social change rather than reaction, was a dramatic about-face for the union's leadership.

In 1973 with Laurel still smoldering, Canadian Keith Johnson was elected president of the International. He was the first Canadian to occupy the presidency since the deportation of Harold Pritchett in 1940, and one of the few Canadians ever to hold that office in any AFL-CIO union.

During the 1970s the IWA renewed its fight for better working conditions and job security for woodworkers. Old problems appeared in new and more sophisticated forms as revolutionary technology and run-away-shops were redefining jobs and closing mills; the new threats to workers' health and safety were sometimes bewilderingly complex; environmental despoliation wrought by corporate clear-cutting blighted entire regions; and at the bargaining table the employers had never been bolder.

Still more profound were the changes occurring in the spheres of corporate organization and social relations. The era of the diversified, multinational corporation was at hand. The Pacific Northwest, although still the operations centre of the IWA, was no longer center stage for a struggle which during the 1960s had moved to the South and in the 1970s had become international in scope; the black movement and the anti-war movement had raised new questions about racism, sexism and corporate colonialism; a new generation of workers questioned the dominance of corporate power over their lives and the authority of their union leaders; young radicals pushed for expanded social and political roles for the union and a reaffirmation of the kind of democratic unionism the IWA had stood for at its founding. In short, as Newfoundland and Laurel had illustrated, the mission of industrial unionism had changed and, as the IWA approached the 1980s, the challenges it faced were no less formidable than those it faced in 1937.

8
Conclusion

Today, the North American trade union movement maintains the status quo arrived at in the decade following World War II. The central events of those years were the ousting of Communist Party leaders in some CIO-CCL unions and the expulsion of other Communist-led unions. Among the conventional explanations for the defeat of the Communists, two of the more compelling ones have been supported through studies of the IWA. The first, put forth by Vernon Jensen, states that the rank-and-file unionists rejected communism and removed Communists through democratic means. A second thesis, introduced by Irving Abella, places major responsibility for the Communists' defeat on themselves. He attributes their defeat to social, tactical, and judgemental errors made at critical junctures. The Jensen and Abella theses both assume that political processes, internal to unions, are the decisive factors in the resolutions of factional differences in those unions.

We have rejected both these theses on empirical and analytical grounds. On the face of it, the Jensen/Abella positions leave too many questions unanswered. For example, if the rank and file was predisposed against communism, why then did the anti-communist blocks in nearly all CIO-CCL unions have to fight pitched battles to put communist exclusion clauses in their constitutions? Why was it necessary for the CIO-CCL to expel the Communist-led unions in the late 1940s rather than simply letting the members of those unions determine who their leaders were going to be? Why was it necessary for the U.S. to spend thousands of dollars on Harold Pritchett's deportation case when a vote of the membership of the IWA could have sent him back to Canada? And was it necessary to ban Communists through the anti-communist clauses of the Taft-Hartley Act if the members of the union were not electing them? The answer, of course, is that the rank and file did continue to elect Communists to leadership positions. In the IWA case, many Communists were elected to office by referendum election as late as the post-World War II years; few were ever defeated by White Bloc opponents; all eventually were purged.

How do we account for the persistence of rank-and-file support for Communists in the IWA? We believe the presence of Communists in the IWA and other CIO-CCL unions provided an alternative to social democratic and business unionism which shifted the balance of class struggle to the advantage of the working class. Moreover, union members recognized the superiority of the Communists' leadership. The results of the IWA's yearly referendum elections confirm this. Between 1936, when Pritchett was elected president of the Federation of Woodworkers, and 1941, when the combined effects of his deportation and the massive intervention of the CIO National Office produced a White Bloc victory, an all-Communist administration had been elected annually by the IWA's membership. Moreover, in 1946, after the CPUSA had, according to the accounts of its critics, discredited itself by supporting the wartime no-strike pledge, two Communist Party members from the Pacific Northwest and one from British Columbia defeated White Bloc incumbents for International office. In British Columbia, the White Bloc was smashed in the referendum elections following the war.

Sociologically, one can explain the persistence of rank-and-file support of the Communist leaders through reference to the primary group relationship between those leaders and the union's members. The Communists had provided necessary leadership during the early years of the crisis and had taken the initiative to organize the industry at a time when no one else was interested. Among lumber workers at the time there was a desperate need for organization, for minimal relief from depression conditions, and for union recognition. The Communists had provided the practical wherewithal to accomplish these objectives; through their work in the Unemployed Councils and the Party's Trade Union Unity League (TUUL) affiliates, they had "delivered the goods."

The real strength of the Communist unionists was that they were indigenous members of the mill towns and logging camps in which they worked and organized. Sometimes immigrants, and sometimes sons and daughters of immigrants, by the mid 1930s, they had, in either case, usually been residents of their communities for about thirty years. Their working class identity derived from their own family and work experience; none had arrived at their radicalism via an academic route; few had even a high school education. IWA Communists learned their politics either from exposure to the IWW movement when they were young or from parents who were socialists.

With this background of a strong primary group bond and a record of solidarity with fellow workers which could be translated into electoral strength in union politics, it is understandable, then, that Pritchett was elected over Hartung in two contests for International President and once for president of the Federation of Woodworkers. Although a favorite in his own Portland district council with its pocket of conservative lumber workers, Hartung was not able to move successfully beyond district-level politics until the CIO National Office appointed him as assistant to Adolph Germer in 1942. With the advantage of his appointed position and a multitude of other changes that occurred in the interim (not the least of which was that all Communists had been banned by the Taft-Hartley Act), Hartung finally achieved his goal of being International President in 1953.

It is significant that Hartung's primary political base was not the rank-and-file woodworkers, but rather the trade union professionals who had preceded him to positions of influence in the CIO. Hartung advanced his career, not on the strength of his personal and family ties to the industry's working class, nor on the strength of his record as an organizer; rather, he found favor with powerful people outside the IWA ranks and advanced when the intervention of the state created the opportunities. In class terms, then, the political base of White Bloc leaders like Hartung was less working class than it was bureaucratic or petty bourgeois.

Communists not only out-performed their opponents as organizers, but they offered a superior concept of what industrial unionism should be. For them, unions were not ends in themselves but rather organizing centers for the working class. One only has to review the newspapers published by the union under the two regimes to get a sense of the difference between them. Under Communist editorship the IWA's paper, the *Timberworker*, opposed clearcutting of forests and log exports and promoted reforestation and conservation. The paper ran regular features on the history of the industry and previous attempts of loggers and mill workers to organize. It published reviews of books and movies of interest to members and printed short stories, poems and songs written by members and authors such as Irene Paull (who often wrote under the name Calamity Jane). Under Communist editorship, the paper carried a full "women's page" devoted to news about the women's auxiliaries and household work. By New Left feminist standards, the content of those pages left much to be desired, but by comparison with the papers published after the White Bloc took over they were remarkably advanced: the White Bloc abolished the women's auxiliaries as part of the purges in the late 1940s. The *Timberworker* carried reports from loggers about camp conditions and letters from local workers who went to Spain to fight in the civil war. If anything, the *Timberworker* was a political paper, since all the issues concerning internal politics were aired in the paper. Indeed, one of the earliest complaints against the Communist leadership was that the paper was *too* political.

Under Communist leadership, the union took positions of the widest possible range on issues affecting lumber workers; its stands on U.S. foreign policy were among the most important. Next to the question of communism in its own ranks, no single issue has divided the American labor movement so much as that of foreign policy. It is not necessary to review here the specific positions taken on the Spanish Civil War, the U.S. entry into World War II, the Marshall Plan, trade with China, or the Vietnam war. The significant point is that the Communists did take positions while their opponents consistently tried to rule foreign policy debates out of order; the left would take straightforward, pro or con positions on the questions, while the White Bloc would try to restrict debates to wages, hour and working conditions.

The authors do not wish to maintain that the contents of the positions taken by the Communists should be exempt from critical examination. The point, however, is that during the period of the CIO-CCL, the *content* was seldom the main issue; the scope of what unionism should be was the issue, and on this subject the Communists clearly had the broadest vision. At the same time, and in anticipation of critics, it must be said that in the IWA's case there is *no* evidence that the *contents* of the foreign policy positions taken by the Communists cost them any strength among the rank and file. Criticisms of the reversals in the Communist Party line were initially made, as nearly as we can tell, by social democrats like Adolph Germer (in the U.S.) and Charlie Millard (in Canada) who represented the CIO and CCL in national affairs; they were not made by rank and file members. It would appear from the evidence in the IWA case that Roger Keeran was correct when he said, in *The Communist Party and the Auto Workers Union,* that the Communist Party's wartime policies probably gained them as many supporters as it lost them.

While there is no evidence in this case that the Communists' position on wartime matters was in any decisive in the purges that followed the war, one can argue that the failure of the U.S. labor movement to consolidate an anti-imperialist position in the post-war years cost workers and unions dearly. At a time when U.S. capital was freely crossing national boundaries to exploit foreign labor, and exporting domestic investment resources and jobs to Europe and the Third World, conservative business unionists like those in the IWA's White Bloc crippled labor's ability to fight back on the same ground. The short-sighted decision to block foreign policy debates proved to be crucial when the international mobility of U.S. capital began sapping the strength of the American working class in the thirty years following the war.

Finally, as union democracy was a major issue for lumber workers, they recognized that the Communists represented a form of unionism that was a clear alternative to the dictatorial form lumber workers had known under the AFL. Although we have not attempted a systematic study of the IWA's organizational structure, it is clear that the Communists stood firmly behind two institutions that are considered litmus tests of union democracy: rank-and-file referendum elections and non-restrictive membership provisions. From the inception of the union under Communist leadership, IWA leaders were elected by yearly referenda. In 1948, when the British Columbia District Council led by Harold Pritchett and others who had for years been Communist leaders in the IWA broke away from the International, the B.C. District reconstituted itself as the Woodworkers Industrial Union of Canada with referendum elections of all officers. Cases like the WIUC are important because they represent Communist unionism in its purest form. A review of examples like the WIUC make it very difficult to sustain the anti-Communist assertion that Communists represented some kind of en-trenched, bureaucratic oligarchy in the CIO-CCL unions.

The protection of open membership provisions was especially important. Divisions between workers have for a long time been the bane of the working class, and formalizing those divisions through exclusions based on race, gender, nation or politics as some unionists have done is irreconcilable with democratic unionism. The Communists' record in the IWA on this question is beyond reproach, but the same cannot be said of their opponents. In 1938, a year after the election of Harold Pritchett's communist administration, the White Bloc began introducing resolutions to deny union membership to Communist Party members. Defeated in 1938, the White Bloc introduced similar resolutions in 1939 and 1940, and each time they were defeated. In 1941, after other factors intervened, a resolution was finally passed which was subsequently used to ban locally elected delegates to conventions and, in one instance, to oust an International vice-President who was only a *former* member of the Communist Party. It is absolutely clear, then, who was denying democratic union rights to whom.

Sides were drawn in the same way in 1947 when the passage of the Taft-Hartley Act required union officials to sign affidavits swearing they were not members of the Communist Party. The White Bloc was quick to take advantage of the law to solidify its own position.

Nevertheless, the Left was eventually ousted from the union and in this book we have attempted to explain the reasons for that ouster. The roots of the political factions involved in this issue were located in the social composition of the industry's workforce which was, in turn, determined by the industry's uneven historical development. The industry established itself in Oregon in the mid-1800s, moved north to Washington toward the end of the century and expanded to British Columbia in the early twentieth century. The structure of the industry in Oregon was characteristically small mills which employed seasonal workers who farmed the rest of the year. Many of these workers were already farmers who had previously settled in the midwest or had come from Europe looking for a chance to survive as small farm owners. Their backgrounds were conservative and their aspirations were bourgeois. The operations in northern Washington and British Columbia, by contrast, were established in the era of monopoly capitalism which allowed for much larger operations and the use of year-round laborers. The large mills concentrated workers in company towns and logging camps, and many of those workers were Scandinavian immigrants who had socialist or trade union experience in Europe.

The different origins of the industry, in turn, help explain the different forms of labor organization. The form of craft unionism that was appropriate for the pre-monopoly industry that began around Portland separated the industry's workers into organizations along their lines of skills or crafts. Some craftsmen and the unskilled workers were left unorganized. When unions seeking to organize all the industry's workers came along, they were viewed by the craft unions as competition. Thus, from the arrival of the IWW in the early 1900s, through the period of the CIO-CCL, and continuing to the present, mill and woods workers in the Pacific Northwest have been divided by competing concepts of unionism. In British Columbia the later development of the industry precluded the early craft union form of organization, and therefore, industrial unionism was established there without a contest.

We have argued that while the industry's workforce was politically divided in the mid to late 1930s, the Communist composition of the IWA's leadership at that time accurately reflected the balance of social forces within the union. In our research we have found no evidence that, in the absence of intervention by the state and the national offices of the CIO and CCL, the rank and file would have rejected its Communist leaders.

Where the departure of CIO-CCL Communists cannot be attributed to a spontaneous anti-communism, the tactical history of the anti-communist campaign becomes a more significant consideration. Use of the organizing program for political purposes and the alteration of the structures of representation within the IWA were perceived by both factions to have been keys to the triumph of the White Bloc. Former International President and White Bloc stalwart James Fadling confirmed this when he spoke before the 1958 convention (see Chapter VII).

This tactical history becomes especially interesting where it intersects with social histories and political-economic studies which attempt to understand the fragmented and uneven development of the North American working class. Conservative and radical rank-and-file politics in the IWA were both determined by the different immigration patterns in the region as well as the uneven development of the industry and the varying impact of logging camps and milltown life on the lumber workers. By premising their organizing priorities on these deeply-rooted historical differences, the White Bloc and Communist organizers provided a subjective component that connected with and amplified the objective condition. By identifying these connecting points, we hope we have made clearly visible the political labyrinth that underlies the CIO-CCL's post-war disintegration, and have helped to prepare the way for a more structural interpretation of their histories.

Appendix

APPENDIX I

RESOLUTION NO. 30A (From the 1941 Convention)

SUBJECT: COMMUNIST PARTY MEMBERS

CONVENTION ACTION: CONCURRED IN.

SUBMITTED BY LOCAL 100 AND LOCAL 11-129.

The Committee recommends that Amendment be amended so that the resolve will read: "Any member accepting membership in the Communist Party shall be expelled from the International Woodworkers of America, and is permanently debarred from holding office in the International Woodworkers of America, and no member of the Communist Party shall be permitted to have membership in our Union, unless they withdraw from the Communist Party and forfeit their membership therein.

"Members charged with membership in the Communist Party shall be tried on said charges as provided for in the Constitution.

"Representations, either verbal or written, of any employer or employer's agent shall not be considered by the Union in determining whether or not anyone charged with being a member of the Communist Party actually holds such membership.

"And that the proposal for a referendum vote on the question be adopted and that the Convention considers the proposal solely on the merit or demerit of submitting to the membership."

APPENDIX II

RESOLUTION NO. 26 (From the 1955 Convention)

SUBJECT: STUDY AND ANALYSIS OF THE STRUCTURE OF THE IW OF A
CONVENTION ACTION: CONCURRENCE AS AMENDED

WHEREAS: The seventeenth Annual Constitutional Convention of the International Woodworkers of America, held in Vancouver, B.C., in 1935, adopted a resolution known as Resolution R-28 calling for the establishment of a large committee composed of the International Executive Board, the International Trustees, the International Officers, the Assistant Director of Organization and the Director of Research for the purpose of analyzing the entire organizational and administrative setup of this International Union; and

WHEREAS: This Committee met and drafted a report for the Eighteenth Annual Constitutional Convention which was originally scheduled for August 1954 in Milwaukee, Wisconsin, but which was postponed until this year — as result of the Northwest lumber strike; and

WHEREAS: The Committee obviously lacked the necessary direction and facts to carry out the intent of Resolution R-28; and

WHEREAS: The recommendation of the Committee did not contain a complete study nor any full and rounded-out plan for improving the efficiency of the entire International Union structure but, in fact, dealt only with one minor phase of the International's operation; now

THEREFORE BE IT RESOLVED: That this Eighteenth and Nineteenth Annual Constitutional Convention of the International Woodworkers of America held in Milwaukee, Wisconsin, August 22–26, 1955, go on record endorsing the intent of Resolution R-28 and that it be more fully developed and implemented than the Committee's report now indicates; and

BE IT RESOLVED: That the Eighteenth and Nineteenth Annual Constitutional Convention of the International Woodworkers of America authorize and instruct the Director of Research, as a member of the Resolution R-28 Committee, to make a full and complete study and analysis of the entire structure of this and other International unions; and

BE IT RESOLVED: That such a study be made free and independent of any individual or group except those whom the Director of Research may designate for assistance; and

BE IT FURTHER RESOLVED: That the cost of making this study be borne by the Emergency Fund; and

BE IT FINALLY RESOLVED: That the Director of Research consider this his principal project, aside from 1956 negotiations, and that he have the study, including cost figures, ready prior to the convening of the Twentieth Constitutional Convention for presentation; — first, to the International Executive Board and the International Trustees; and, second to the Twentieth Annual Constitutional Convention of the International Woodworkers of America when it convenes.

SUBMITTED BY: Klamath Basin District Council, No. 6, IWA-CIO.

s/ Tim Sullivan, President

s/ H.E. Geiger, Secretary-Treasurer

APPENDIX III

Portland, Oregon
October 26, 1951

Mr. Fadling:

And now you drink and dine with the Commies, solicit and accept their support to win an election. They can have you, Mister.

The Denver Convention brings to a climax a long series of activities on your part which only hard facts could positively convince of. A book could be filled with an account of them, and each item would be as solidly and factually true as the very singular truth I shall bother to deal with herein.

I will confine comment to the present. You are right now cuddled arm in arm with KARLY LARSON as he half-carries you up a political hill which it is obvious now you lacked the self-stature to climb on the power of ability. Pushing hard from the rear is a flock of lesser Commie rats whose very guts I hate, as do thousands of other bona fide union men. The membership elevated you for your fight against scummy rats in the past. Today they are your bedfellows. From here on out, I will consider it an honor to be found high on your list of enemies.

I have made my judgment of the intra union dispute centered around you, on *facts* only. I have been patient and given you every break in arriving at an unemotional and factual opinion. Now I'll speak that opinion, and with the emotion of disgust and contempt. People *throughout* the International can actually produce factual proof of book length covering practices of yours that might all well be added up to total this question: How little can a little man be?

I fully expect the combination of FADLING-BACKED-BY-LARSON to win the election now at issue. KARLY LARSON, YOUR BUDDY, might represent a world-wide movement that is *very* effective in winning elections. The same crowd is also very effectively killing American boys in Korea while the American comrades are dining and drinking with "friends." I need not tell you, for you well know, that you CANNOT win without the support of your Commie dinner guests. *If* you win, it will be by the margin of votes the Commies control or manipulate. *That,* you have long known. It will be their hands that hang the Victory Medal on your chest, that of a wholly incompetent and very little man. Only the most shriveled personality could receive their awards, for they can reach no higher than the gutter's edge. I spent years fighting to build this union, and then fighting the old alignment of KARLY LARSON, MAURICE RAPPORT, HAROLD PRITCHETT, etc., and now a new alignment forms. Cashing in what principle you might have had, you have bought yourself a presidency, MAYBE, — BUT, you have bought yourself a REAL fight with a multitude of men of principle — and, Mister, I'll be there.

/s/ Don Helmick
Don Helmick
Member IWA Local No. 5-3

SOURCE: Letter found in Adolph Germer papers at University of Oregon Library.

Notes

Introduction

1 The major works done on the International Woodworkers of America are Vernon Jensen, *Lumber and Labor* (New York: Farrar & Rinehart, 1945); Irving Abella, *Nationalism, Communism, and Canadian Labour* (Toronto: University of Toronto Press, 1973). Chapters or sections of books treating the IWA can be found in Irving Bernstein, *Turbulent Years* (Boston: Houghton Mifflin, 1970); Walter Galenson, *The CIO Challenge to the AFL* (Cambridge: Harvard University Press, 1960). Myrtle Bergren's *Tough Timber* (Toronto: Progress Books, 1966) is a good first-hand account of the organizing of Vancouver Island written by the wife of one of the organizers. "Official" versions of IWA history can be found in Grant MacNeil's *The IWA in British Columbia* (Vancouver, British Columbia: Regional Council No. 1 International Woodworkers of America, 1971) and International Woodworkers of America, *History of the International Woodworkers of America* (Portland, Oregon: IWA, 1974). For valuable insights on British Columbia see also: Paul Phillips, *No Power Greater, A Century of Labour*

in B.C. (Vancouver: B.C. Federation of Labour, 1967) and Gad Horowitz, *Canadian Labour in Politics* (Toronto: University of Toronto Press, 1968).
2 Jensen, *Lumber and Labor*, pp. 269-70.
3 Abella, *Nationalism.* p. 138.

Chapter 1

1 Robert Pike, *Tall Trees, Tough Men* (New York: W.W. Norton & Company, Inc., 1967), pp. 1-51.
2 Vernon Jensen, *Lumber and Labor* (New York: Farrar & Rinehart, 1945), p. 45. Isaac Stephenson, "Recollections of A Long Life, 1829-1915" (Chicago: Self-published, 1915).
3 Pike, *Tall Trees*, p. 90.
4 Jensen, *Lumber and Labour*, p. 36.
5 Ibid., p. 51.
6 Ibid., p. 53.
7 Ibid., pp. 54-55.
8 Ibid., p. 57.
9 Marvin Dunn, "Kinship and Class: A Study of the Weyerhaeuser Family." (Ph.D. Dissertation, University of Oregon, 1977), p. 74.
10 Jensen, *Lumber and Labor*, p. 63.
11 Dunn, "Kinship and Class," p. 182.

12 This description of the industry in the Southern States is taken from Jensen, *Lumber and Labor*, pp. 71–94.

13 Thomas R. Cox, *Mills and Markets: A History of the Pacific Coast Lumber Industry to 1900* (Seattle: University of Washington Press, 1974), p. 9.

14 Ibid., p. 25.

15 Murray Morgan, *The Last Wilderness* (Seattle: University of Washington Press, 1955), pp. 60–63.

16 Cox, *Mills and Markets*, p. 118.

17 Ibid., p. 134.

18 For more complete discussions of the role of immigrants in shaping the politics of the North American working class, see Stanley Aronowitz, *False Promises* (New York: McGraw-Hill, 1973), pp. 137–213; David Brody, *Steelworkers in America* (New York: Harper, 1960); Melvyn Dubofsky, *Industrialism and the American Worker* (Arlington Heights: AHM, 1974); Philip Foner, *The Industrial Workers of the World, 1905-1917* (New York: New World Paperbacks, 1965); Al Gedicks, "The Social Origins of Radicalism among Finnish Immigrants in Midwest Mining Communities," Review of Radical Political Economics 8(3), Fall, 1976; Herbert Gutman, *Work, Culture and Society* (New York: Vintage, 1977); Sam Kushner, *Long Road to Delano* (New York: International Publishers, 1975); Gerald Rosenblum, *Immigrant Workers* New York: Basic Books, 1973); Carlos A. Schwantes, *Radical Heritage: Labor, Socialism, and Reform in Washington and British Columbia, 1885-1917* (Seattle: University of Washington Press, 1979).

19 Jurgen Kuczynski, *The Rise of the Working Class* (New York: World University Library, 1967), p. 145.

20 Paul Phillips, *No Power Greater* (Vancouver: B.C. Federation of Labour, 1967), pp. 25, 45–46.

21 Kuczysnki, *The Rise of the Working Class*, pp. 211–212.

22 The Germans did make a contribution to the overall labor politics of British Columbia and Washington. Schevantes notes the presence there of the Socialist Labor Party, which had heavy German involvement prior to 1900.

23 John Lindberg, *Background of Swedish Emigration in the United States* (Minneapolis: University of Minnesota Press, 1930), p. 135.

24 Ibid., pp. 207–222.

25 BC Overtime, "Logger's Navy – History of the IWA in British Columbia" (Vancouver: BC Overtime – Educational Radio Productions, 1976). A copy of this tape has been made available to the authors by Howie Smith of BC Overtime.

26 Schwantes, *Radical Heritage*, p. 16.

27 Ibid., p. 26.

28 Robert Christie, *Empire in Wood* (Ithaca: New York State School of Industrial and Labor Relations, 1956), p. 196.

29 Vernon Jensen, *Lumber and Labor*, p. 119.

30 Robert Tyler, *Rebels of the Woods: The IWW in the Pacific Northwest* (Eugene: University of Oregon Books, 1967), pp. 8, 23, 26, 212.

31 The origin of the name "Wobblies" is uncertain. One version quotes a letter written by Wobbly Mortimer Downing. "'In Vancouver, in 1911, we had a number of Chinese members, and one restaurant keeper would trust any member for means. He could not pronounce the letter 'w,' but called it 'wobble' and would ask: 'You I Wobble Wobble?' and when the card was shown credit was unlimited. Thereafter the laughing term amongst us was 'I Wobbly Wobbly,' and when Herman Suhr, during the Wheatfield Strikes, wired for all footloose Wobblies to hurry there, of course the prosecution made a mountain of mystery out of it, and the term has stuck to us ever since . . ." Quoted in Jack Scott, "How the Wobblies Got Their Name," in his *Plunderbund and Proletariat* (Vancouver, B.C.: North Star Books,

1975), p. 153.

32 Jensen, *Lumber and Labor*, pp. 119–120.

33 Tom Scribner, "Lumberjack" (Manuscript dated 1966, in the Oregon Collection, University of Oregon Library), p. 2. Scribner, a former Wobbly and Communist Party member, has written a series of essays about union struggles he participated in and assembled them in manuscript form.

34 Jensen, *Lumber and Labor*, p. 127.

35 Melvyn Dubofsky, *We Shall Be All* (New York: Quadrangle, 1969), p. 413.

36 Jensen, *Lumber and Labor*, p. 135.

37 Richard Boyer and Herbert Morais, *Labor's Untold Story,* New York, United Electrical Workers, 1973, p. 212.

38 Ralph Chaplin, *The Centralia Conspiracy* (Chicago: Charles H. Kerr, 1973), p. 66.

39 Jensen, *Lumber and Labor*, p. 142; also Chaplin, *The Centralia Conspiracy*, pp. 68–80; and the "Centralia Defence Committee" papers, and "Free Ray Becker Committee" papers at the Oregon Historical Society, Portland, Oregon.

40 Tyler, *Rebels of the Woods*, p. 214.

Chapter 2

1 Charlotte Todes, *Labor and Lumber* (New York: International Publishers, 1931), p. 102.

2 Unemployment Insurance Commission of the State of Oregon, *Report and Recommendations* (Salem: State of Oregon, 1935), p. 54.

3 *Ibid.*, pp. 44, 69, 78.

4 Vernon Jensen, *Lumber and Labor* (New York: Farrar and Rinehart, Inc., 1945), p. 151.

5 Al Hartung Memoirs.

6 Jensen, *Lumber and Labor*, p. 151.

7 *Ibid.,* p. 137.

8 William Z. Foster, *American Trade Unionism* (New York: International Publishers, 1947), pp. 200-201. Irving Bernstein, *The Lean Years* (Boston: Houghton Mifflin Company, 1960), pp. 139-141. "Boring from within" was a popular phrase used to distinguish the strategy of activists working for reforms within established AFL unions, from the strategy of building new unions (known as "dula unionism"). The phrase was never intended to imply that the Communist presence in the unions was any less organic.

9 *Voice of Action*, July 6, 1934, p. 4. The *Voice of Action* was a left-wing labor paper published in Seattle, beginning in March, 1933, that carried extensive news about unemployment demonstrations and organizations in the Pacific Northwest. The paper also circulated widely in Portland, and in 1934 a branch office was opened in the city to facilitate distribution of the paper.

10 Todes, *Labor and Lumber*, p. 188.

11 Interviews with Ralph Nelson by Jerry Lembcke, 1976, Portland.

12 Henry Buechel, "Labor Relations in the West Coast Lumber Industry: 1935" (unpublished Master's thesis, Department of Economics, State College of Washington, 1936), p. 29.

13 *Voice of Action*, December 19, 1933, p. 1.

14 Dave Yorke, "The Worker's Unity League in B.C." (unpublished Bachelor's thesis, Department of History, Simon Fraser University, 1973), pp. 17-25.

15 The sketch of the 1931 Fraser Mills Strike and Harold Pritchett is drawn primarily from William Tattam's 1967, 1972 and 1978 interviews with Pritchett; a paper by Pritchett, "Maillardville-Coquitlam Ratepayers," March, 1978, in author's possession. Other useful interviews with Pritchett are by Clay Perry, October 27-31, 1978, Harold Pritchett Collection, University of British Columbia, Vancouver, B.C., (hereafter cited as "Harold Pritchett Collection"); Harold Wynn, October 31, 1970, "Harold Pritchett Papers," University of Washington Library, Seattle, Washington, and by students at Simon

Fraser University, June, 1975, Simon Fraser University, Vancouver, British Columbia. See also: Chris Potter, "Leading B.C. Labor Figure tells how it is–and was," *The Herald*, January 21, 1975, p. 13; and Fred Wilson, "A Woodworker's Story," *Pacific Tribune*, April 29, 1977, pp. 11-13.

16 *Pacific Tribune*, April 19, 1977, p. 12.

17 David Stone, "The IWA: The Red Bloc & White Bloc." Burnaby, B.C.: Simon Fraser University, 1973, p. 3.

18 Interview with Harold Pritchett, October 27, 1978, by Clay Perry. "Harold Pritchett Collection."

19 Interview with Harold Pritchett, October 27, 1978 by Clay Perry; interview with Harold Pritchett, November 11, 1978, by the authors; and *Pacific Tribune*, April 29, 1977, pages 12 and 13.

20 See papers, strike bulletins, etc., for Fraser Mills Strike. Uncatalogued. "Harold Pritchett Collection."

21 *The B.C. Lumber Worker*, Volume 1, Number 1, November 7, 1931, p. 4. "Harold Pritchett Collection."

22 In 1928, 9.33% of the total work population in British Columbia was Asian, declining to 8.28% by 1934. In the lumber industry the comparable figures were 18.2% in 1929 and 15% in 1934. British Columbia Department of Labour, *Annual Report* (British Columbia: British Columbia Department of Labour, 1934), pp. 15, 20, 27 and 29 "cited by" L.T. Smythe, "The Lumber and Sawmill Worker's Union in British Columbia." (unpublished Master's thesis, Department of History, University of Washington, 1937).

23 See various issues of *The B.C. Lumber Worker*. "Harold Pritchett Collection."

24 Myrtle Bergren, *Tough Timber: The Loggers of B.C. — Their Story* (Toronto: Progress Books, 1966), pp. 55-56.

25 *Strike Bulletin* (Lumber Worker's Industrial Union. February, 1934). "Harold Pritchett Collection." Yorke, "The Workers' Unity League," p. 21.

26 Bergren, *Tough Timber*, pp. 47-52; Yorke, "The Workers Unity League," p. 22.

27 Interview with Harold Pritchett by Clay Perry, October 27, 1978. "Harold Pritchett Collection." Bergren, *Tough Timber*, pp. 36-37.

28 Bergren, *Tough Timber*, p. 38. Yorke, "The Workers' Unity League," p. 21.

29 *Local 1-80 Bulletin Centennial Edition* (Local 1-80: Duncan, B.C. 1971), p. 4.

30 Bergren, *Tough Timber*, p. 52.

31 *Ibid*

32 *Portland Oregonian*, February 27, 1930, p. 5.

33 *Ibid.*, February 11, 1931, p. 1.

34 Thomas Graham, "Administration of Unemployment Relief in Oregon, 1930-1936" (unpublished Bachelor's thesis, Department of History, Reed College, June 1936, pp. 76-77; Wayne W. Parrish and Wayne Weishaar, *Men Without Money, The Challenge of Barter and Scrip* (New York: G.P. Putnam's Sons, 1933), pp. 68-69; Arthur M. Schlesinger, Jr., *The Crisis of the Old Order, 1919-1933* (Boston: Houghton Mifflin Company, 1957), p. 251; Robert Tyler, *Rebels of the Woods: The IWW in the Pacific Northwest* (Eugene: University of Oregon Books, 1967), pp. 218-222.

35 *Oregon Labor Press*, December 8, 1933, p. 4; January 5, 1934, p. 4.

36 *Ibid.*, November 17, 1933, p. 4.

37 Michael Edwin Thompson, "The Challenge of Unionization; Pacific Northwest Lumber Workers During the Depression," (unpublished Master's thesis, Department of History, Washington State University, 1968), p. 21.

38 Interview with Julia Ruuttila, June 24, 1968, Portland, Oregon by William Tattam. Julia Ruuttila's husband participated in the attempts to organize Portland sawmill workers in the early 1930s. By 1933 she was actively engaged in local unemployment demonstrations. From 1934 to 1941, she helped organize demonstrations in support of the 1934 Longshore Strike, the 1935 Lumber Strike, the unemployed,

the CIO movement, and such left-wing peace movements as the 1937 Emergency Peace Mobilization meeting in Chicago.

39 *Portland Oregonian*, May 10, 1934, p. 1.

40 Jensen, *Lumber and Labor,* p. 160; Ralph Nelson Interview; *Oregonian,* May 13, 1934, p. 14; Thompson, "The Challenge of Unionization," p. 46.

41 The sketch of activities in San Francisco is drawn mainly from Larrowe's Chapter 3, "Where Did You Study Military Strategy and Tactics," pp. 62-94. Charles Larrowe, *Harry Bridges: The Rise and Fall of Radical Labor in the United States* (New York: Lawrence Hill, 1972).

42 *Portland Oregonian*, July 14, 1934, p. 1.

43 Samuel Yellen, *American Labor Struggles* (New York: S.A. Russell, 1956), p. 354; *Portland Oregonian*, July 20, 1934, p. 1.

44 The 1919 Oregon Criminal Syndicalism Act classified three types of conduct as felonious: first, advocating or teaching criminal syndicalism, orally or by printing; second, organizing a society which advocated it; third, taking part in conducting a meeting of such a society. The prohibited acts were punishable by imprisonment for not less than one year nor more than 10 years, or a fine of not more than $1,000, or both.

45 M. Paul Holsinger, "Patriotism and The Curbing of Oregon's Radicals, 1919–1937," (A Paper Presented to the Seventeenth Annual Conference of the Western History Association, Portland, Oregon, October 14, 1977), pp. 5-7; *Oregonian*, July 19, 1934, p. 1; July 22, 1934, p. 1.

46 *Voice of Action*, November 23, 1934, p. 1; *Portland Oregonian*, November 23, 1934, p. 1.

47 Cathy Howard, "The Case of Dirk DeJonge," *Portland Oregonian*, "Northwest Magazine," March 28, 1976, p. 9.

48 Portland Central Labor Council, Papers and Correspondence, Box 5. Oregon Historical Society, Portland, Oregon.

49 Henry Buechel, "Labor Relations in the West Coast Lumber Industry," (M.A. Thesis, State College of Washington,

1935), pp. 43, 55.

50 Jensen, *Lumber and Labor*, pp. 153-159. For a detailed description of the NRA Lumber Code, see Francis Clare Vause, "The Administration of the National Industrial Recovery Act With Particular Reference to the Lumber Industry" (unpublished Bachelor's thesis, Department of History, Reed College, Portland, Oregon, 1934).

51 Thompson, "The Challenge of Unionization: Pacific Northwest Lumber Workers During the Depression," p. 43.

52 Claude W. Nichols, Jr., "Brotherhood in the Woods: The Loyal Legion of Loggers and Lumbermen, a Twenty Year Attempt at Industrial Cooperation" (unpublished Ph.D. dissertation, University of Oregon, 1959), pp. 168-174. Robert W. Tyler, "The United States Government as Union Organizer: The Loyal Legion of Loggers and Lumbermen," Mississippi Valley Historical Review, XLVII (December, 1960), pp. 443-451. The Wagner Act also contributed to the decline of the 4-L by prohibiting employer contributions to the organization. The Loyal Legion responded by removing employers as members, adopted the possibility of striking, and changed its name to the Industrial Employees Union. But the I.E.U. could not compete with the AFL and the fast-developing CIO and slowly withered away. (See Tyler, "The United States Government as Union Organizer," p. 451.)

53 Robert Christie, *Empire in Wood* (Ithaca, New York: Cornell University Press, 1956), pp. 288, 289, 290.

54 *Ibid.*, p. 289.

55 *Voice of Action*, April 12, 1935, p. 1.

56 Blaine Crawford, "Activities of the CIO and AFL in the Pacific Northwest Lumber Industry, 1935-1940" (unpublished Master's thesis, Department of Economics, University of Idaho, 1942), p. 17.

57 *Portland Oregonian,* May 16, 1935, p. 1; N. Sparks, "The Northwest General Strike," *The Communist*, September, 1935.

58 Philip Foner, The Industrial Workers of the World, 1905-1917 (New York: New World Paperbacks, 1973), pp. 221-25; Walter Mattila, "Finnish Paul Bunyans" (Portland: Finnish American Historical Society of the West, 1962), pp. 14-15; Jeremy R. Egolf, "The Limits of Shop Floor Struggle: Workers and the Bedaux System at Willapa Harbor Lumber Mills, 1933–1935." Paper presented to the Midwest Sociological Society Annual Meeting, April 7-9, 1982, Des Moines, Iowa.

59 Egolf, "The Limits of Shop Floor Struggle," p. 10; In 1938, Charles Bedaux was appointed head of the commercial operations of the German industrial complex of I.G. Farber although he was an American citizen at the time. In 1942, the U.S. government jailed Bedaux for collaboration with the Nazis; Charles Higham, *Trading with the Enemy*, New York, Delacourt Press, page 178.

60 *Ibid*, p. 2, 20.

61 *Ibid.*, pp. 21-22.

62 *Ibid.*, p. 3.

63 *Portland Oregonian,* May 11, 1935, p. 1; May 12, 1935, p. 1.

64 *Voice of Action,* May 14, 1935, p. 1.

65 *Ibid,* June 7, 1935, p. 1; *Portland Oregonian*, June 5, 1935, p. 4.

66 Buechel, "Labor Relations in the West Coast Lumber Industry," p. 84.

67 Walter Galenson, *The CIO Challenge to the AFL* (Cambridge: Harvard University Press, 1960), p. 320; Max Kampelman, *The Communist Party vs. the CIO: A Study in Power Politics* (New York: F.A. Praeger, 1957), p. 12; David Saposs, *Communism in American Unions* (New York: McGraw Hill, 1969), p. 15.

68 Buechel, "Labor Relations in the West Coast Lumber Industry," p. 30.

69 *Voice of Action,* February 23, 1935, p. 3.

70 *Ibid.*, April 26, 1935, p. 2; Interview with Al Hartung by William Tattam, June 4, 1968 at Portland, Oregon.

71 Interview with Karly Larsen by Jerry Lembcke, c. 1978 at Stanwood, Washington.

72 William Tattam, "Sawmill Workers and Radicalism: Portland, Oregon 1929–1941" (Unpublished Master's thesis, Department of History, University of Oregon, 1979), p. 56, 57; Paul William Parks, "Labor Relations in the Grays Harbor Lumber Industry." (Unpublished Master's thesis, Department of History, University of Washington, 1948), p. 31; Interview with Ralph Nelson by Jerry Lembcke, c. 1976 at Portland, Oregon.

73 Interview with Lyman Wax by Jerry Lembcke, October, 1978 at Sheridan, Oregon.

74 Buechel, "Labor Relations in the West Coast Lumber Industry," p. 55.

75 *Voice of Action*, May 14, 1935, p. 1; *Oregon Labor Press*, June 14, 1935, p. 1.

76 Vernon Jensen, "Labor Relations in the Douglas Fir Lumber Industry" (Unpublished Ph.D. Dissertation, University of California, 1939), p. 231; *The Oregonian,* June 7, 1935, p. 1; Letter from G.H. Bottwell, Klickitat Fir Lumber Company, to Charles H. Martin, Governor of Oregon, June 17, 1935, Charles H. Martin Papers, Oregon Historical Society.

77 Statement by Orville Smith, President of the Northwest Joint Strike Committee, before a Committee meeting in Portland, Oregon. *Portland Oregonian*, June 15, 1935, p. 2; *Voice of Action*, June 14, 1935, p. 2.

78 Interview with Ralph Nelson by Jerry Lembcke, 1976 at Portland, Oregon; Interview with Harry Pilcher by William Tattam, June 12, 1968 at Portland, Oregon; Buechel, "Labor Relations in the West Coast Lumber Industry," p. 73.

79 Buechel, "Labor Relations in the West Coast Lumber Industry," pp. 170-175 for samples of C.C. Crow's Strike Bulletins.

80 For this section, see the previously quoted works from Henry Buechel and Michael Thompson. Selden C. Menefee's "Tacoma, Timber and Tear Gas," *Nation* (July 17, 1935), pp. 76-77, is helpful. Frank Onstine's "The Great Lumber Strike in Humboldt County, 1935" (self-

published paper, Eureka, undated, but probably late 1970s) is an excellent manuscript. See also Tattam's "Sawmill Workers and Radicalism..." pp. 50-64. The strike stimulated two fine "proleterian" novels of the 1930s which have been overlooked. See *Disillusion, A Story of the Labor Struggle in the Western Wood-Working Mills*, by Ben. H. Cochrane and William Dean Coldiron, (Portland: Benfords and Mort, 1939) and *Marching, Marching* by Clara Weatherwax, (New York: John Day, 1935).

81 *Portland Oregonian*, June 8, 1935, p. 2.

82 Buechel, "Labor Relations in the West Coast Lumber Industry, p. 92.

83 *Seattle Post-Intelligencer*, June 18, 1935, p. 1, "Cited by" Buechel, p. 88.

84 Tattam, "Sawmill Workers and Radicalism," p. 60.

85 Buechel, "Labor Relations in the West Coast Lumber Industry," p. 90.

86 Jensen, *Lumber and Labor*, pp. 203-204; *Voice of Action*, October 25, 1935, p. 3.

87 *B.C. Lumber Worker*, April 11, 1936, p. 2.

88 Bergren, *Tough Timber*, p. 96; Interview with Harold Pritchett by Jerry Lembcke and William Tattam, November 11, 1978 at Port Coquitlam, British Columbia.

89 Minutes and Resolutions of the Sawmill and Timber Worker's District Councils Joint Executive Committee Conference." (Centralia, Washington, June 27-28, 1936). In authors' possession.

90 Jensen, *Lumber and Labor*, p. 204.

91 *Ibid.*, September 20, 1936, p. 41.

92 Walter Galenson, *The CIO Challenge to the AFL* (Cambridge: Harvard University Press, 1959), p. 384.

93 United Brotherhood of Carpenters and Joiners of America, *Proceedings of the Twenty-Third General Convention* (Lakeland, Florida, December 7-15, 1936), pp. 37-38.

94 Galenson, *The CIO Challenge to the AFL*, p. 385. Christie, *Empire in Wood*, p. 298.

95 Federation of Woodworkers, *Proceedings of the First General Convention* (Portland, Oregon, September 18-22, 1936), pp. 4-8.

96 Federation of Woodworkers, *Proceedings of the Second Semi-Annual Convention* (Longview, Washington, February 20-22, 1937), p. 4. "Cited by" Jensen, p. 206 and Galenson, *The CIO Challenge*, p. 385.

97 *Ibid.*, September 19, 1936, p. 15; September 20, 1936, pp. 36, 40. The convention also went on record in support of the immediate release of Ray Becker, IWW member still imprisoned for events in Centralia, Washington, 1919, plus freedom for Tom Mooney and the Scottsboro Boys. Pritchett later traveled to the state penitentiary in Walla Walla, Washington, to visit Becker and present a copy of the convention's resolution to him and also visited with Tom Mooney at San Quentin prison in California.

Chapter 3

1 Jensen, *Lumber and Labor*, pp. 189-90.

2 Taped interviews with Irene Paull by Linda Stiles, April 6 and 17 and May 5 and 9, 1979, San Francisco, California. Irene Paull was involved with the Midwest loggers' union movement from its beginning in 1936. A reporter for the Duluth *Timber Worker* and editor of *Midwest Labor*, her pen name "Calamity Jane" was well-known to the region's loggers and sawmill workers. Until his untimely death in 1944, she was married to Hank Paull, attorney for the IWA. Tape 7. Debra E. Bernhardt, "We Knew Different: The Michigan Timber Workers' Strike of 1937" (Unpublished Master's Thesis, Department of History, Wayne State University, 1977), p. 3. The authors are grateful to Ms. Bernhardt for her permission to use and quote from this most useful thesis.

3 Bernhardt, "We Knew Different: The Michigan Timber Workers' Strike of 1937," p. 3; *Midwest Labor*, July 22, 1938, p. 2.

4 Interview with Ernest Tomberg by William Tattam, June 6, 1979. Mr. Tomberg was active in the Minnesota lumber workers' struggle to form a union. In 1937 he was the business manager of the Duluth *Timber Worker*; from 1937–1939 he was an IWA executive board member and was an executive officer in the Midwest District Council of the IWA until 1947. Currently he works in labor relations for Georgia-Pacific.

5 Bernhardt, "We Knew Different: The Michigan Timber Workers' Strike of 1937," p. 3.

6 *Ibid.*, p. 10.

7 *Timber Worker* (Duluth), May 28, 1937.

8 Bernhardt, "We Knew Different: The Michigan Timber Workers' Strike of 1937," p. 14.

9 Bernhardt, "We Knew Different," p. 19; Interview with Oliver Rasmussen, 1978.

10 Bernhardt, "We Knew Different," pp. 24-29. See also Jensen, *Lumber and Labor*, pp. 197-198; National Labor Relations Board, *Decisions and Orders*, Vol. xvii, November, 1939, pp. 799-800.

11 Linda Stiles Interview with Irene Paull, May 5, 1979, Tape 7, side 1.

12 *Ibid.*, *Midwest Labor*, March 18, 1938, p. 6. See also Bernhardt, "We Knew Different," pp. 31-33 and Irene Paull, "Farewell, Sweet Warrior, To My Husband, Henry Paull, Attorney for the People," May, 1944. Irene Paull Collection, unprocessed, Minnesota Historical Society, St. Paul, Minnesota. "Cited by" Bernhardt, p. 32.

13 Interview with George Rahkonen, strike activist, by Debra Bernhardt, Phelps, Wisconsin, July 29, 1977. "Cited by" Bernhardt, p. 35. Bernhardt, "We Knew Different," pp. 40-42. *Midwest Labor*, March 18, 1938, p. 6.

14 Letter of Henry Paull to Nathaniel Clark, July 4, 1937. Frank Murphy Collection, Box 18, Folder 1, Michigan Historical Collections, Bentley Library, Ann Arbor, Michigan. "Cited by" Bernhardt, "We Knew Different," p. 43.

15 For the activities the night of June 30, see Bernhardt, pp. 42-44; Jensen, *Lumber and Labor*, p. 199; *Midwest Labor*, March 18, 1938, p. 16, and Linda Stiles interview with Irene Paull, May 5, 1979, tape 7.

16 Interview with Irene Paull by Linda Stiles, May 5, 1979, tape 7, side 2.

17 Bernhardt, "We Knew Different," pp. 46-53.

18 *Milwaukee Journal*, June 6, 1937. "Cited by" Bernhardt, p. 32.

19 *Galenson*, The CIO Challenge, p. 386; *Oregonian*, June 8, 1937, p. 1.

20 Interview with Al Hartung by Wm. Tattam. *Timberworker,* June 11, 1937, p. 1; *Portland Oregonian*, June 19, 1937, p. 1. The Columbia River District Council controlled loggers and sawmill workers on the Oregon side of the Columbia River from the Dalles to Astoria, and down the Oregon coast to Coos Bay.

21 Interview with Al Hartung. Hartung was present at the meeting; Interview with Harold Pritchett.

22 *Portland Oregon Journal*, July 8, 1937, p. 6.

23 Carlie May Gilstrap, "A Study of the Labor Dispute and the Various Decisions Involved in the Controversy at the Plylock Plant and the M & M Woodworking Company" (unpublished Master's thesis, Department of Economics, University of Oregon, June, 1940), p. 17.

24 Portland *Oregon Journal*, July 19, 1937, p. 2.

25 Al Hartung Memoirs (unpublished), copies in authors' possession. Interview with Don Helmick by Wm. Tattam, June 5, 1968, at Portland, Oregon. Don Helmick had been active in the Sawmill and Timber Workers Union, AFL, and was an executive officer in the Federation of Woodworkers. From 1937 to 1941, he was an executive board member of the Columbia River District Council. *Oregonian*, July 25, 1937, p. 1.

26 Margret Glock; *Collective Bargaining in the Pacific Northwest Lumber Industry* (Berkeley: Institute of Industrial Rela-

tions, 1955), p. 13.

27 *Portland Oregonian*, July 27, 1937, p. 1.

28 American Federation of Labor, *Report of the Proceedings of the 57th Annual Convention* (Denver, Colorado, 1937), p. 468.

29 *Portland Oregonian*, August 17, 1937, p. 1.

30 Glock, *Collective Bargaining*, p. 13.

31 *Portland Oregonian*, August 20, 1937, p. 1; August 17, 1937, p. 1.

32 Weinstein, Jacob Joseph, "The Jurisdictional Dispute in the Northwest Lumber Industry," (B.A. Thesis, Reed College, 1939), pp. 53-54.

33 *Portland Oregonian*, December 10, 1937, p. 1.

34 Thompson, "The Challenge of Unionization: Pacific Northwest Lumber Workers During the Depression," p. 100.

35 Glock, *Collective Bargaining*, p. 15.

36 *Portland Oregonian*, July 7, 1939, p. 1; April 2, 1943, p. 1.

37 Al Hartung Memoirs, pp. 26-27.

38 *Portland Oregonian*, September 3, 1937, p. 10.

39 Interview with Lyman Wax. Al Hartung Memoirs, p. 30.

40 Al Hartung Memoirs, pp. 44. The comment on the Bridges-Hartung relationship was mentioned by Stanley Earl at a retirement testimonial for Al Hartung in 1967. At that time Earl was a Portland City Commissioner, but in 1937 he was a business agent for the Portland IWA local.

41 Thompson, "The Challenge of Unionization," pp. 98-106.

42 *New York Times*, December 19, 1937, p. 22. "Cited by" Thompson, p. 109.

43 *Portland Oregonian*, March 4, 1938, p. 3; April 29, 1939, p. 1; Jensen, *Lumber and Labor*, p. 221.

44 National Lawyers Guild, *Report of the Civil Liberties Committee* (Portland: Oregon Chapter, May, 1938), pp. 6-7, Oregon Historical Society.

45 *Ibid.*, p. 11; Walter B. Odale, editor, *Weekly Report of Communist Activities*, January 1, 1937, p. 2, Charles H. Martin Papers, Oregon Historical Society. (Mimeographed)

46 Interview with James Fanz, June 18, 1968 by William Tattam, Portland; Interview with Al Hartung, 1968 by William Tattam, Portland.

47 *Weekly Report of Communist Activities*, February 19, 1937.

48 *Portland Oregonian*, November 28, 1937, Section VI, p. 3.

49 Americans Incorporated, *Radical Activities Bulletin*. 1938-1939. Various issues contained in the "Charles H. Martin Papers," Oregon Historical Society.

50 Letter from Americans Incorporated to Gust Anderson, secretary of the Central Labor Council, February 14, 1939. Central Labor Council Papers, Box 10, Oregon Historical Society. Nelson Hibbs, leader of Americans Incorporated, was an ex-Naval officer and Naval Aide to Oregon Governor Martin.

51 Harry Whitten, "Subversive Activites: Their Extent and Meaning in Portland," (unpublished Bachelor's thesis, Department of History, Reed College, 1939), pp. 131-132; *Oregonian*, November 21, 1937, p. 3; National Lawyers Guild, *Report of the Civil Liberties Committee*, pp. 427-437 and p. 45.

52 *Portland Oregonian*, November 28, 1937, Section VI, p. 3.

53 *Labor New Dealer*, December 18, 1937, p. 6; Whitten, "Subversive Activities," p. 138.

54 See Charles H. Martin Papers, Portland Police Bureau File, Oregon Historical Society. The Report was titled "Additional Information, Labor Trouble and Communists," and written on an Oregon State Police Officers Report Form.

55 Central Labor Council, *Minutes and Proceedings*, April 26, 1937, p. 1; February 7, 1938, p. 417.

56 Telegram from Gust Anderson, secretary of the Central Labor Council, to Cordell Hull, Secretary of State, September 24, 1937. Central Labor Council Papers, Box 10, Oregon Historical Society.

57 Portland *Oregon Journal*, September 21,

1937, p. 1.

58 *Ibid.*

59 Letter from the Portland Central Labor Council to Hon. Walter M. Pierce, U.S. Representative, containing resolution unanimously endorsed by the CLC, March 14, 1938, censoring the union activities of Harold Pritchett. Pierce Papers, Oregon Historical Collection, University of Oregon.

60 Telegram from R.J. Dant of Dant and Russell, Inc., to Walter M. Pierce, U.S. Representative, February 17, 1938. Pierce Papers, Oregon Historical Society.

61 International Woodworkers of America, *Proceedings of the Third Constitutional Convention* (Klamath Falls, Oregon, October, 1939), p. 20.

62 *Investigations of Un-American Propaganda Activites in the U.S.* Hearings before the Special Committee on Un-American Activities, House of Representatives, 76th Congress, 1st Session, 1939, pp. 1717, 2014-2015, 2910-2911.

63 *Ibid.*, pp. 2014-2015, 2910-2911.

64 International Woodworkers of America, *Proceedings of the First Constitutional Convention* (Portland, Oregon, 1937), pp. 111-112.

65 *Ibid.*, p. 144.

66 *Portland Oregonian*, December 8, 1937, p. 6. International Woodworkers of America, *Proceedings of the First Constitutional Convention,* pp. 144-146.

67 *Timberworker*, December 11, 1937, p. 11; *Portland Oregonian*, December 8, 1937, p. 6. In a similar strike at Toledo, Oregon, Portland longshoremen awaited the withdrawal of IWA pickets before starting to load the *Anna Schafer* with lumber produced at the C.P. Johnson Co. where the IWA had called a strike five months earlier. Eventually, three IWA organizers from the Coos Bay (IWA) Council were found guilty of contempt of court for refusing to comply with a restraining order which would have stopped picketing and allowed the longshoremen to load the lumber. *Portland Oregonian,* December

16, 1937, p. 1; December 24, 1937, p. 1.

68 International Woodworkers of America, *Proceedings of the First Constitutional Convention,* p. 151.

69 International Woodworkers of America, *Proceedings of the Second Constitutional Convention* (Seattle, Washington, 1939), pp. 116, 124.

70 *Ibid.*, p. 256.

71 Interview with Harold Pritchett by William Tattam, March 20, 1972; A 39-page history of the struggle between the Left and Right issued March 20, 1940, by the IWA administration, contains numerous reprints of newspaper articles, summaries of meetings, and copies of letters, proceedings and hearing which are helpful. Harold Pritchett Papers, University of Washington Library, Seattle, Washington.

72 "Conversations That Took Place Between Joe Clark, James 'Red' Fadling, John Deskins, and William Anderson," May 21, 1939. Transcribed recording made by Ernie Dalskog. Authors' possession.

73 *Ibid.,* Parks, p. 47; Jensen, *Lumber and Labor,* p. 232.

74 Aberdeen *World,* October 24, 1939.

75 International Woodworkers of America, *Proceedings of the Third Constitutional Convention* (Klamath Falls, Oregon, 1939), p. 76 Klamath Falls *Evening Herald* October 19, 1939.

76 International Woodworkers of America, *Proceedings of the Third Constitutional Convention,* pp. 76, 79, 81, 104. Walter Galenson, *The CIO Challenge to the AFL* (Cambridge: Harvard University Press, 1960), p. 396. Portland *Oregon Journal,* October 19, 1939, p. 6.

77 International Woodworkers of America, *Proceedings of the Third Constitutional Convention,* pp. 79, 11; Portland *Oregonian,* October 20, 1939, p. 1.

78 International Woodworkers of America, *Proceedings of the Third Constitutional Convention,* p. 95; Interview with Harold Pritchett by Clay Perry, October 27, 1978, "Harold Pritchett Collection;"

Interviews with Harold Pritchett by Clay Perry, October 31, 1978, "Harold Pritchett Collection," and by William Tattam, May 25, 1968.

79 International Woodworkers of America, *Proceedings of the Third Constitutional Convention,* pp. 137, 288.

80 *Ibid.,* Interview with Harold Pritchett by William Tattam, May 25, 1968; Interview with Harry Bridges by William Tattam, June 10, 1968, San Francisco; *Timberworker,* "The Disruptionists — An Enemy Within Our Walls," April 29, 1939, p. 4 and "Expose the IWA Wreckers," June 3, 1939, p. 1.

81 International Woodworkers of America, *Proceedings of the Third Constitutional Convention,* p. 305.

82 Portland *Oregon Journal,* October 22, 1939, p. 1; Hartung Interview; International Woodworkers of America, *Proceedings of the Third Constitutional Convention,* p. 305.

83 The Finnish Hall history is drawn primarily from Parks, "Labor Relations in the Grays Harbor Lumber Industry," pp. 54-72.

84 Aberdeen *World,* December 1, 1939, p. 1. "Cited by" Parks, p. 66.

85 Parks, "Labor Relations in the Grays Harbor Lumber Industry," p. 69.

86 The account of the Laura Law murder is based primarily on Irvin Goodman and John Caughlan's "The Truth in the Case of Laura Law, Murdered Labor Leader," (October, 1950) Goodman and Caughlan were attorneys for the Law family and represented Dick Law at the hearings following the murder. The large, 11-page pamphlet quoted extensively from testimony at the coroner's inquest and was issued ten years after the case in an attempt to clarify the lingering doubt about who killed Laura Law. (Copy in author's possession). See also Parks, "Labor Relations in the Grays Harbor Lumber Industry," pp. 75-88 for a good account of the murder drawn from primary sources.

87 Parks, "Labor Relations in the Grays Harbor Lumber Industry," p. 82. *Crow's Pacific Coast Lumber Digest,* "What Are We Waiting For;" An editorial by C.C. Crow, November 15, 1939. "Cited by" Goodman and Caughlan, "The Truth in the Case of Laura Law," p. 10.

88 Goodman and Coughlan, "The Truth in the Case of Laura Law," p. 6-10.

89 Seattle *Times,* June 4, 1940, p. 1. "Cited by," Parks, "Labor Relations in the Grays Harbor Lumber Industry," p. 88.

90 That the murder of Laura Law had lasting effects on the IWA's development is the opinion of Dick Law's attorney, John Caughlan. Interview with John Caughlan by Jerry Lembcke, November 26, 1978, Seattle, Washington. Interest in the Laura Law case was revived in 1983 when police records became available for the first time. According to Tom Churchill, who is researching the materials for a novel on Laura Law, the collection contains depositions taken from the principles in the case which were never seen by Dick Law or his attorneys. The collection is available at the Washington State Archives in Olympia.

Chapter 4

1 Michael Widman to John L. Lewis, June 3, 1940, Adolph Germer Papers; "History of the IWA Administration," pp. 18-23, Harold Pritchett Papers, University of Washington Library.

2 *Labor New Dealer,* January 12, 1940, p. 1; *Timberworker,* January 20, 1940, p. 1; Jerry Lembcke and William Tattam, "The International Woodworkers of America: A Challenge to Vernon Jensen's Thesis of Rank and File Anti-Communism," (unpublished paper presented to the Twelfth Annual Pacific Northwest Labor History Conference, May 4-6, 1979, Spokane, Washington), p. 13.

3 *Labor New Dealer,* January 26, 1940, p. 1, January 19, 1940, p. 1; Minutes of

the IWA Executive Board Meeting, January 11-12, 1940, Klamath Falls, Oregon.

4 Interview with Harold Pritchett by William Tattam, May 25, 1968; Notice to Pritchett from the U.S. Department of Labor Immigration and Naturalization, Vancouver, B.C., April 27, 1937 and July 30, 1937, Harold Pritchett Papers; Paul Coughlan, Pritchett's lawyer, to Representative Charles H. Learly, House of Representatives, Washington, D.C., November 5, 1937, p. 3, Harold Pritchett Papers.

5 The pamphlet issued by the Committee is included in the Tom Burns Papers, AX 108, Box 1, Oregon Historical Collection, University of Oregon.

6 Houghton, Chuck, and Coughlan, Pritchett's lawyers, to The Honorable Charles H. Leary, House of Representatives, Washington, D.C., November 5, 1937, Harold Pritchett Papers.

7 Letter from Lee Pressman to Harold Pritchett, February 2, 1938. Harold Pritchett Papers.

8 Interview with Harold Pritchett by William Tattam, March 20, 1972.

9 Interview with Lyman Wax, September 28, 1978, at Sheridan, Oregon, by Jerry Lembcke. Mr. Wax was a prominent member of the opposition in the CRDC.

10 Warren Magnuson to Paul Coughlan, December 9, 1937. Harold Pritchett Papers.

11 Wayne Morse to Mickey Orton, August 17, 1940. Harold Pritchett Papers.

12 Wayne Morse to Cordell Hull, August 17, 1940. Harold Pritchett Papers.

13 Paul R. Josselyn, American Consul General, Vancouver, B.C., to Harold Pritchett, August 22, 1940. Harold Pritchett Papers.

14 Harold Pritchett to O.M. Orton. (undated) In authors' possession.

15 Interview with Harold Pritchett by William Tattam, March 20, 1972. Brief of Harold Pritchett to the U.S. Department of Justice: Immigration and Naturalization Service, Vancouver, British Columbia, March 12, 1948, p. 15. (In authors' possession).

16 Harvey A. Levenstein, *Communism, Anti-communism, and the CIO* (Westport: Greenwood Press, 1981), p. 266.

17 *Ibid.*, p. 89. Lewis' relationship with the Communists took many forms during the CIO years. Whether he was personally behind the anti-Left campaign at this particular point is a matter of some controversy.

18 James Prickett, for example, argues, "Disdain for anti-communism, coupled with firm guarantees of autonomy for each affiliate, characterized the CIO until the post-war period. But at the 1946 convention, the seemingly inexplicable occurred: anti-communism was endorsed and autonomy severely infringed..." See Prickett's "Some Aspects of the Communist Controversy in the CIO," *Science and Society*, 33 (3) (Summer-Fall, 1969), p. 302.

19 Widman to Lewis, June 3, 1940. Adolph Germer's personal papers are in the Wisconsin Historical Society. Cited hereafter as Adolph Germer Papers.

20 Len DeCaux, *Labor Radical* (Boston: Beacon Press, 1970), p. 462; Interview with Harold Pritchett by Jerry Lembcke, July 2, 1976: Lorin Lee Cary, "Adolph Germer: From Labor Agitator to Labor Professional" (Ph.D. Dissertation, University of Wisconsin, 1968), pp. iii-iv; B.Y. Mikhailov, et. al., *Recent History of the Labor Movement in the United States* (Moscow: Progress, 1977), p. 53.

21 An interview with Bill Harris by Jerry Lembcke, Reedsport, Oregon, 1977.

22 Interview with Bill Harris by Jerry Lembcke, 1977, Reedsport, Oregon; Interview with Carl Winn by Jerry Lembcke (telephone); Al Hartung, "Unpublished Memoirs."

23 Pritchett to District Councils and Locals, June 19, 1940. Adolph Germer Papers.

24 Germer to John Gibson, July 13, 1940, Adolph Germer Papers.

25 Pritchett's list, attached to Germer to Widman August 23, 1940, recommended

N.E. Mason (Coeur d'Alene), William Gettings (Eureka), Fred Lum (Vernonia), Richard Law (Aberdeen), John Sullivan (Portland), James Freeman (North Bend, Washington), William Riley (Bessemer, Alabama), Ernest Tomberg (Duluth, Minnesota), and Steve Severson (Portland).

At the August 9 meeting, Worth Lowery supported: Max Gardiner (Wuana), Jodie Eggers (Carlton), Harvey Nelson (Molalla), Lawrence Daggett (Astoria), Claude Ballard (Astoria), Clarence Monroe (Beaver Creek), K.S. Paddock (Port Oxford), A.F. Benedict (Clallan Bay).

Orton and McCarty supported: Addie Pearson (Coeur d'Alene), N.E. Mason (Coeur d'Alene), William Gettings (Eureka), and John Maki (Michigan). See also Germer Memo, August 9, 1940. Adolph Germer Papers.

26 Exactly how Germer gained information on the political leanings of individuals isn't clear. His correspondence does indicate, however, that he had a network of informants around the country. A letter to Gwen Lurie in New York City on July 23, 1940, for example, read: "Enclosed you will find a folder which contains a list of 'sponsors' of the 'mobilize for Peace' meeting in Chicago, August 31 to September 2. I wish you would go over the list and check the names of those you know to be 'labeled.' There are some that I know. There are many others I do not–and you have ways of finding out those things."

27 Germer to Haywood, August 19, 1940, Adolph Germer Papers. Germer to Basil Hoke, August 16, 1940, Adolph Germer Papers.

28 Pritchett to Germer, August, 1940. Adolph Germer Papers.

29 Germer to Haywood, August 22, 1940, Adolph Germer Papers. When Germer learned that "suddenly and mysteriously a number of allegedly CPs or fellow travelers [were] appearing in the district where we intend to place organizers," he noted with pleasure that Lee Hall and Gola Whitlow had been assigned to him from the National and requested again that Harvey Fremming, Freddie Thomason and others be sent. "In view of the activities of the CP and others," Germer wrote to Haywood, "our work will be rendered more difficult and I know you will give us all available help."

30 Widman to Germer, September 9, 1940, Adolph Germer Papers.

31 Congress of Industrial Organizations, "Record of Hearing conducted by CIO-appointed Committee of the Controversy Between Designated International Officers of the International Woodworkers of America and Adolph Germer, CIO-appointed IWA Organizational Director." Seattle: Congress of Industrial Organizations, February 5, 1941), p. 1147. See testimony of Francis Murnane.

32 Germer to Haywood, September 25, 1940; Francis to J.C. Lewis, December 16, 1940, Adolph Germer Papers.

33 John L. Lewis to Germer, September 26, 1940, Adolph Germer Papers.

34 *Ibid.*; Interview with Brick Mohr by Jerry Lembcke, January 14, 1978. Mr. Mohr was a well-known logger who worked extensively in the Aberdeen area.

35 International Woodworkers of America, *Proceedings of the Fourth Constitutional Convention* (Aberdeen, Washington, 1940), p. 14.

36 *Ibid.*, p. 287. Other handbills, bearing a red hammer and sickle and the words "The Commies are Coming; The Commies are Here," had been distributed in Aberdeen and Hoquiam before the IWA Convention opened. Union officials maintained that the handbills were an attempt by the AFL to discredit the convention. Two men, one an AFL organizer, were later detained by local police for distributing the handbills. *Oregonian*, October 8, 1940, p. 1; *Oregon Journal*, October 8, 1940, p. 6.

37 International Woodworkers of America, *Proceedings of the Fourth Constitutional Convention*, 1940, p. 14.

38 International Woodworkers of America,

Proceedings of the Fourth Constitutional Convention, p. 38.

39 International Woodworkers of America, *Proceedings of the Fourth Constitutional Convention*, p. 109.

40 Lembcke and Tattam, "The International Woodworkers of America," p. 29.

41 *Ibid.*

42 International Woodworkers of America, *Proceedings of the Fourth Constitutional Convention*, p. 184. Germer to Haywood, October 19, 1940; Germer to Haywood, October 13, 1940; O.M. Orton to John L. Lewis, October 11, 1940; John L. Lewis to O.M. Orton, October 14, 1940. Adolph Germer Papers.

43 Portland *Oregon Journal*, October 13, 1940, p. 1. Interview with Al Hartung by William Tattam.

44 *Ibid.*

45 Portland *Oregon Journal*, October 18, 1940,. p. 1; Haywood to Germer, October 21, 1940, Adolph Germer Papers.

46 Jensen, *Lumber and Labor,* p. 239, Don Latvalla to Germer, September 9, 1940. Germer Papers. Congress of Industrial Organizations, "Record of Hearing..." February 5, 1941, pp. 1353-55.

47 *Ibid.*, January 31, 1941, pp. 550-51.

48 The district councils involved were the Northern Washington, Grays-Willapa Harbors, Columbia River, Plywood and Veneer, and Boommen and Rafters. See Jensen, *Lumber and Labor*, pp. 246-49.

49 Jensen, *Lumber and Labor*, pp. 249-50.

50 *Ibid.*, p. 255.

51 Brick Mohr, interview, January 14, 1978; Congress of Industrial Organizations, "Record of Hearing...," January 31, 1941, pp. 540-42; Germer to Haywood, March 19, 1941.

52 Fremming to Coulter, December 19, 1940, Germer Papers; Galenson, *The CIO Challenge to the AFL*, p. 401.

53 *Timberworker*, December 14, 1940, p. 1.

54 Portland *Oregonian*, December 7, 1940, p. 2.

55 *Timberworker*, December 7, 1940, p. 1;

December 21, 1940, p. 1.

56 The proceedings against Germer were actually convened as a hearing. A quasi-legal format was used, however, and the IWA referred to the proceedings as a trial. The record of the trial is contained in nine volumes of transcribed proceedings entitled, "Record of Hearing conducted by CIO appointed Committee of the Controversy Between Designated International Officers of the International Woodworkers of America and Adolph Germer, CIO-appointed IWA Organizational Director," dated January 28, 1941–February 6, 1941. Eight of the volumes are located in the Internatonal Office of the IWA, Portland, Oregon and one volume for February 4, 1941 was found in Adolph Germer's papers at the Wisconsin Historical Society, Madison, Wisconsin. A copy of that volume is in the possession of the authors.

57 "Record of Hearing," January 29, pp. 18-22.

58 *Ibid.*, January 29, p. 22; January 31, pp. 545-46, 566, 644.

59 Mitch and Cowherd to Germer, September 6, 1940. "We recommend Riley..." "...as far as T.H. Beaird is concerned, he is not a very strong man. He is more along the lines of a poor local preacher, with about 90 pounds of energy to put forth his program."; Germer to Cowherd and Mitch, Septemer 9, 1940, Germer refused to hire Riley and asked for another recommendation; Germer to Haywood, September 25, 1940, Cowherd and Mitch recommended Oscar Pruett whom Germer also refused to hire. Germer ruled out several people recommended as organizers by the IWA on similar grounds. He didn't want leaders elected by the rank and file to be in charge of the organizing program because of the likelihood they would be leftists. The IWA argued that by putting people in the field who were already in elected positions, the organizing program would more likely be able to speak to the needs of workers and that the

union could save money by having fewer people on salaries. It was pointed out to Germer that Beaird was also on the IWA payroll when he was hired as an organizer, making a strong case against Germer for discrimination. Cowherd and Mitch to Germer, October 14, 1940.

60 "Record of Hearing," January 29, pp. 184-218.

61 *Ibid.*, January 29, pp. 99-102; January 30, pp. 301, 304-09.

62 Germer to Widman, August 9, 1940; Germer to Haywood, October 17, 1940; Germer to Haywood, July 1, 1941; Dalyrmple to Murray, December 20, 1940.

63 That Germer's primary mission as Director of Organization of the IWA was the breaking of the left wing political control of CIO unions on the West Coast was suggested to the authors by labor economist and historian Irving Richter and by Fred Gadroury, editor of *Labor Today*.

64 Germer to Haywood, October 13, 1940.

65 "Record of Hearing," February 6, p. 1524.

66 Congressman Frank Fries to Germer, October 16, 1940, Adolph Germer papers.

67 The other two were D.L. Shamley and H.I. Tucker "Record of Hearing," February 6, pp. 839, 956, 1053.

68. *Ibid.*, February 4, 1941, pp. 819-25.

69 Lorin Lee Cary, *Adolph Germer: From Labor Agitator to Labor Professional* (Ph.D. Dissertation, University of Wisconsin, 1968).

70 "Record of Hearing," February 3, 1941, p. 684.

71 *Ibid.*, p. 924.

72 *Ibid.*, p. 923.

73 *Ibid.*, p. 949.

74 "Record of Hearing," February 4, p. 951.

75 *Ibid.*, p. 1008.

76 "Record of Hearing," February 6, pp. 1527-34.

77 *Ibid.*, pp. 1540-43.

78 Cary, *Adolph Germer*, pp. 152-53, 162.

79 Portland *Oregonian*, January 31, 1941, p. 10.

80 Al Hartung, Interview, June 4, 1968; *Oregonian*, February 1, 1941, p. 1; *Oregon Journal*, February 3, 1941, Section II, p. 1; Tattam, Chapter VI, pp. 158-59.

81 Letter from Joseph B. McAllister, President of the PIUC to John L. Lewis, President, Congress of Industrial Organizations, October 28, 1940. Joseph B. McAllister, Miscellaneous Papers, Oregon Historical Collection, University of Oregon.

82 *Ibid.*

83 Portland *Oregon Journal*, February 3, 1941, Section II, p. 1.

84 Nelson to Curran, March 13, 1941, Adolph Germer Papers; McCarty to Haywood, March 2, 1941; Germer to Cowherd, March 11, 1941; Galenson, *The CIO Challenge*, p. 402.

85 "Dick" to Germer, March 15, 1941, Adolph Germer Papers. Irving Richter suggests that "Dick" may have been Dick Deverall.

86 C. Schwantes, *Radical Heritage, Labor Socialism and Reform in Washington and British Columbia, 1885–1917* (University of Washington Press, 1979); Jerry Lembcke, "The International Woodworkers of America" (Ph.D. Dissertation, University of Oregon, 1978), Chapter VI.

87 Germer to Haywood, April 1, 1941, Adolph Germer papers; Ron Roley, Interview, June 23, 1977.

88 International Woodworkers of America, *Proceedings of the Fifth Constitutional Convention* (Everett, Washington, October 8-13, 1941), pp. 74-75.

89 Germer to Haywood, April 10, 1941: "In the Southern Washington District Council, which denied Lowery the floor last Friday, there is another sample of Communist democracy. Each local has a member on the Executive Board. Each member on the Board has one vote no matter how large or small his local is. Aberdeen has over 3,000 members. Their delegates have one vote, and another local with 20 or 30 members also has one vote." Adolph Germer Papers.

90 Jerry Lembcke, "The International Woodworkers of America"(Ph.D. Dissertation, University of Oregon, 1978), Chapter V.
91 Galenson, *CIO Challenge*, p. 405.
92 Adolph Germer Diary, August 7, 1941. Adolph Germer Papers.
93 International Woodworkers of America, *Proceedings of the Fifth Constitutional Convention*, p. 194.
94 Adolph Germer Diary, October 8, 1941. Adolph Germer Papers.
95 Lembcke, "The International Woodworkers of America," Chapter V. These figures are taken from International convention roll-call vote records for 1940-42.
96 Interview with Ellery Foster, October 17, 1981, at Winona, Minnesota, by Jerry Lembcke.

Chapter 5

1 For the comparative study of B.C. and the Pacific Northwest, see Jerry Lembcke, "The International Woodworkers of America" (Ph.D. Dissertation, University of Oregon, 1978), Chapter IV. The IWA case is confirmation of Harvey Levenstein's point that the absence of craft unions in some industries made it easier for industrial unions to take root during the 1930s. See Harvey A. Levenstein, *Communism, Anticommunism, and the CIO* (Westport: Greenwood Press, 1981), p. 59. For other histories of the IWA in British Columbia, see Grant MacNeil, *The IWA in British Columbia* (Vancouver: Regional Council No. I, International Woodworkers of America, 1971), pp. 35-36; Paul Phillips, *No Power Greater* (Vancouver: British Columbia Federation of Labour, 1967), p. 116.
2 Myrtle Bergren, *Tough Timber* (Toronto: Progress Books, 1966), p. 136.
3 *Ibid.*, p. 193.
4 *Ibid.*. p. 208.
5 *Ibid.*, p. 215.

6 *Ibid.*, p. 216.
7 *Proceedings of Investigation...CIO*, 1945, Vol. II, pp. 272-87.
8 *Ibid.*, p. 281 and leaflets that were entered as exhibits.
9 *Ibid.*, p. 67.
10 See Harold Pritchett, "Maillardville..."
11 *Proceedings of Investigation...CIO*, 1945, Vol. II, pp. 395; in the exhibits see the "Union Facts" leaflets.
12 See *Canadian Woodworker*, Vol. I, No. 3, p. 2; Minutes of White Bloc Meeting Held September 25-27, 1949.
13 For the involvement of Ballard see *Proceedings of Investigation...CIO*, 1945, Vol. III, p. 363; for the involvement of Brown see Vol. II, pp. 176-77.
   It isn't exactly clear why the Fraser Mills workers were receptive to the appeals of the anti-communist dissidents. Mill workers were generally conceded to be more conservative than loggers, and indeed the respective organizing programs of the White Bloc and Red Bloc were premised upon that fact (see, for example, the division over that question in other parts of this book.) The conservative effect of the company town paternalism and presence of the Catholic Church in the lives of Fraser Mills workers is probably an additional important consideration. Still, the left-wing Workers Unity League had successfully countered the church's efforts to win the allegiance of the workers during the 1931 Fraser Mills strike.
14 Gad Horowitz, *Canadian Labour in Politics* (Toronto: University of Toronto Press, 1968).
15 Quoted by Horowitz, *Ibid.*, p. 67.
16 *Ibid.*, p. 87.
17 *Ibid.*, p. 87.
18 Chairman, CCF Trades Committee to B.J. Melsness, February 4, 1944. District Exhibit #27 in *Proceedings of Investigation...CIO*, 1945.
19 Whelan was associated with a Trotskyist faction within the CCF. He later be-

came a leader in the B.C. Teamsters Union. The relationship between the Trotskyist movement and the CCF, and the role of the Trotskyists within the IWA is addressed again in footnote 83.

20 Horowitz, *Canadian Labour in Politics,* pp. 119-20.

21 Fergus McKean to Provincial Council of the CCF, September 4, 1943; Secretary, CCF, to Fergus McKean, September 13, 1943; see leaflet "The CCF and the Labor-Progressives" by C. Grant MacNeil, M.L.A. Angus MacInnis Collection, University of British Columbia Special Collections Library.

22 Horowitz, *Canadian Labour in Politics,* p. 120.

23 *Ibid.*, p. 120.

24 Ulinder to Kierstad, November 2, 1944. Letter located in *Proceedings, Trial of John Ulinder.*

25 *Proceedings, Trial of John Ulinder,* pp. 1-7.

26 *Ibid.*, pp. 1-2.

27 Members of the jury were: Lorne Atchison, Scotty Sutherland, E. Whalley, J. Malbon, Geo. Hauk, B. Schofield, J. Stewart, Ben Farkes, E. Sandborg, K. Thornley, R. Yates.

28 *Proceedings, Trial of John Ulinder,* pp. 6-7. Olkovich named L. Whalen as the chairman of the meeting and Bill Kierstead, Robertson, Ulinder, Watson, Charles Widen and Tom Barnett as persons attending.

29 *Ibid.*, p. 8.

30 Olkovich, *Statement of Fred Olkovich,* p. 2. This written statement is part of the trial documents.

31 Agitation on the no-strike policy "was one of the chief ways in which the group hoped to drive a wedge between rank and file members and the Union leaders." Olkovich, *Statement by Fred Olkovich,* p. 4.

32 "It was finally agreed, after a discussion, that since many of the [Malaspina] group's supporters had recently joined the Union if they suddenly left other members might believe that they had joined only to cause disruption. Consequently the proposed policy of having the group's supporters drop out of the Union was abandoned." Olkovich, *ibid.*, p. 5.

33 International Woodworkers of America, *International Executive Council Proceedings* (Portland: International Woodworkers of America, May 9, 1945), pp. 25-69; International Woodworkers of America, *Proceedings of District Council No. 1 Executive Board,* March 7, 1945 (Vancouver: International Woodworkers of America, 1945).

34 International Woodworkers of America, *Proceedings of International Executive Board Meeting,* May 9, 1945, pp. 22-35.

35 *Ibid.*, pp. 66-67.

36 See the leaflets entitled "The Undercut" and "Union Facts" in the exhibits collection of *Proceedings of Investigation Committee CIO,* 1945, exhibits 30, 31, and 32.

37 International Woodworkers of America, *Proceedings of International Executive Board Meeting,* August 5, 1945, pp. 47-64.

38 *Ibid.*, pp. 50-51.

39 *Ibid.*, p. 57.

40 *Ibid.*, pp. 63-64. On August 12, 1945, T.G. MacKenzie, Vice President of Local 1-367, recommended that sub-locals be abolished and locals be chartered in each area of the Fraser Valley: Hammond, Mission, Harrison Mills, Chilliwack, Hope and Harrison Lake. That plan never reached fruition.

41 Greenall continued to be an important figure in the IWA. In 1946 his election as International Trustee signaled the resurgence of the left in the International. In 1948 he was expelled from that office under the provisions of the U.S. Taft-Hartley Law. His expulsion was the precipitating event of the "breakaway." Greenall has written a political analysis of those years, *The IWA Fiasco.*

42 *Proceedings of Investigation Committee CIO*, 1945, Vol. I, p. 31.

43 *Ibid.*, Vol. III, p. 315.

44 International Woodworkers of America, *Proceedings of International Executive Board Meeting,* May 11, 1945.

45 International Woodworkers of America, *Special Executive Council Meeting of District Council 1,* April 25, 1945, p. 1.

46 *Proceedings of International Executive Board Meeting,* May 11, 1945, p. 71.

47 IWA, *Special Executive Council Meeting of District Council 1,* April 25, 1945, p. 4.

48 IWA, *Proceedings of International Executive Board Meeting,* May 11, 1945, p. 100.

49 Mitchell had been implicated in the anti-communist "Old Timers" group in the New Westminster local.

50 IWA, *Proceedings of International Executive Board Meeting,* May 11, 1945, p. 93.

51 *Ibid.*, pp. 98-101; IWA, *Proceedings of International Executive Board Meeting,* August 2-4, 1945, p. 40.

52 IWA, *Special Executive Council Meeting of District Council 1, May 20, 1945,* p. 15.

53 *Ibid.*, p. 2.

54 *Ibid.*, p. 11.

55 All during the 1940s the British Columbia Communists were hampered by U.S. immigration officials. Nigel Morgan and Ernie Dalskog, who were both on the Board, were stopped numerous times and prevented from attending important functions of the International. The most important incident of harassment was at the time of the 1947 International convention at which the Taft-Hartley Act was discussed. The Canadian Communist delegates were unable to attend although they were ultimately banned from holding office by its provisions. For a more lengthy treatment of this issue, see Jerry Lembcke. "International Woodworkers of America..." and Lembcke, "Labor Radicalism and the State in the Pacific Northwest — The International Woodworkers of America (CIO)" (Paper pre-sented to the Twenty-Eighth Annual Meeting of the Society for the Study of Social Problems, San Francisco, September 1, 1978).

56 IWA, *Proceedings of International Executive Board Meeting,* August 2-4, 1945, p. 40. The allegation that McKenzie passed the letter to Brown was made during the CIO investigation of the controversy over the organizing program. See *Proceedings of Investigation...CIO,* 1945, Vol. II, p. 237. 242-45.

57 See Al Parkin's history of the IWA in *The B.C. Lumber Worker,* June 2, 1947, p. 6; June 22, 1946, p. 1.

58 *B.C. Lumber Worker,* June 22, 1946, p. 1.

59 *Ibid.*

60 See Harold Pritchett's account of the 1946 strike in *The B.C. Lumber Worker,* August 8, 1946, p. 6.

61 Germer to Eugene Patton, March 4, 1946, in the Adolph Germer Papers (copies in authors' possession).

62 Richard O. Boyer and Herbert Morais *Labor's Untold Story* (New York: United Electrical, Radio & Machine Workers of America 1973), p. 345.

63 Chamber of Commerce of the United States," Communists Within the Labor Movement" (Washington, D.C.: Chamber of Commerce of the United States, 1947), pp. 25, 42.

64 International Woodworkers of America, International Executive Board Minutes, July 22-23, 1947, pp. 28-38.

65 Philip Murray's letter is reproduced in International Woodworkers of America, Internatonal Executive Board Minutes, August 21-30, 1947, p. 13.

66 See Jack Greenall, *The IWA Fiasco* (Vancouver: Progressive Workers Movement, 1965), p. 9. Greenall's work appears to be from a Maoist perspective. He is very critical of communist "revisionism" and of communist leaders like Pritchett and Larsen for having "sold out."

67 International Woodworkers of America, *Proceedings of the Eleventh Constitution-*

al Convention (St. Louis, Mo., August 26-29, 1947), pp. 262-263.

68 *Ibid.*, p. 259.

69 *Ibid.*, pp. 104-105, 110-111.

70 *Ibid.*, p. 117.

71 *Ibid.*, pp. 109-128.

72 Fadling to Larsen and Laux, September 18, 1947; Larsen and Laux to Fadling, September 19, 1947. Both letters are reprinted in the *B.C. Lumberworker*, September 22, 1947.

Today Larsen defends his resignation on the grounds that "we had no way to mobilize ranks in the field against the International and government." (Karly Larsen, Interview, February 23, 1977).

Jack Greenall (*IWA Fiasco*, p. 6) calls Larsen "the undisputed leader of all the progressive forces within the IWA."

73 Joseph Starobin, *American Communism in Crisis, 1943-1957* (Berkeley: University of California Press, 1975), p. 169.

74 See *Daily Peoples World*, April 29, 1949, p. 3; sworn affidavit signed by member of IWA Local 7-140 on March 24, 1949.

75 International Woodworkers of America, International Executive Board Minutes, November 18, 1947, supplements 2-4.

76 *B.C. Lumber Worker*, January 28, 1948, p. 8; *B.C. Lumber Worker*, January 12, 1948, p. 1; International Woodworkers of America, International Executive Board Proceedings, March 9-10, 1948.

77 Riddell, Stead, Graham & Hutchinson to the B.C. District Council, May 12, 1948; Emil Bjarnason to Jerry Lembcke, June 22, 1981; Irving Abella, *Nationalism, Communism, and Canadian Labour,* p. 130; David Stone, "The IWA: The Red Bloc and White Bloc."

78 Trustees of B.C. District Council No. 1, "Untitled Report," June 21, 1948; Abella, *Nationalism,* p. 130.

79 Transcript of "The Voice of the International Woodworkers," January 26, 1948. Copy in possession of the authors.

80 *Ibid.*, August 9, 1948; Interview with Ellery Foster by Jerry Lembcke, October 17, 1981 at Winona, Minnesota.

81 Lorin Lee Cary, "Adolph Germer: From Labor Agitator to Labor Professional," Ph.D. Dissertation, University of Wisconsin, 1968, p. 163; International Woodworkers of America, International Executive Board Proceedings (March 9, 1948), pp. 17-20.

82 Abella, *Nationalism*, p. 117, 118.

83 *Ibid.*, pp. 117-120. The role played by Trotskyists in the campaign against the B.C. District has never been clarified. In 1946 Lloyd Whalen who was an active leader in the anti-communist campaign left the CCF in favour of a Trotskyist faction with two other prominent CCF organizers, R.W. Bullock and T.J. Bradley. According to Ruth Bullock, Mr. Bullock was "pulled out of the shipyards in 1946-47" to "defeat the Stalinist stranglehold on the IWA." Additional documentation on the relationship between the CCF and Trotskyists can be found in the Angus MacInnis Memorial Collection, University of British Columbia Special Collections Library.

84 *Ibid.*, pp. 125-26.

85 *Ibid.*, pp. 130-32; John Ball, Interview, October, 1976.

86 Abella, *Nationalism,* p. 134.

87 Previous accounts of the WIUC have been sketchy, at best. The availability of the proceedings of the disaffiliation convention, the first WIUC convention and the WIUC newspapers make possible the first detailed account of the WIUC's history. District officers at the time were Ernie Dalskog, Harold Pritchett, Hjalmer Bergren, Mark Mosher, and Jack Forbes. *Proceedings*, October 3, 1948, Appendix #1.

88 *Proceedings,* October 3, 1948, p. 32.

89 *Ibid.*, pp. 32, 43.

90 *Ibid.*, pp. 46, 49, 59.

91 *Ibid.*, p. 62.

92 *The Canadian Woodworker,* November 3, 1948, p. 1.

93 *Proceedings,* October 3, 1948, p. 79.

94 *Ibid.*, pp. 87-89.

95 The first WIUC convention was held

October 23 and 24, 1948, and was attended by 244 credentialed delegates, 41 fraternal auxiliary delegates and 48 visitors. Local 71 with fifty delegates had the largest representation.

96 Woodworkers' Industrial Union of Canada, *Proceedings of Constituent Convention*, October 23-24, 1948.

97 *The Canadian Woodworker*, Vol. I, No. 1, p. 1; Vol. I, No. 2, p. 3.

98 *Proceedings*, First Constitutional Convention, p. 43.

99 *Ibid.*, pp. 5, 12, 28.

100 *Ibid.*, pp. 40-42; Supplement #14.

101 *Ibid.*, pp. 8, 13, 16-18, 30, 36, 49; Supplement #14.

102 *Ibid.*, Supplement #14; *The Canadian Woodworker*, Vol. 1, No. 1, p. 8.

103 Joe Morris, "Communism in the Trade Unions" (unpublished speech delivered at Queens University, Circa 1965, copy in the possession of the authors), p. 44.

104 International Woodworkers of America, International Executive Board Proceedings, February 15, 1949, Supplement #3.

105 *The Canadian Woodworker*, Vol. I, No. 2, p. 1; see also Joe Morris, "Communism in the Trade Unions."

106 *The Canadian Woodworker*, Vol. I, No. 4, p. 4.

107 Morris, "Communism in the Trade Unions," pp. 47-48.

108 *The Canadian Woodworker*, Vol. I, No. 4, p. 1.

109 The five strikers charged were Mike Farkas, George Stevens, Otto MacDonald, Alex Armella and Lang Mackie; *The Canadian Woodworker*, Vol. I, No. 6, p. 1.

110 *The Canadian Woodworker*, Vol. I, No. 5, p. 1; Vol. I, No. 4, p. 1; Vol. I, No. 9, p. 1; Vol. I, No. 7, p. 1.

111 Copies of these papers are in possession of the authors.

112 *The Canadian Woodworker*, Vol. I, No. 4, p. 3; Vol. I, No. 7, p. 1; Vol. I, No. 12, p. 1; International Woodworkers of America, International Executive

Board Proceedings, 15 February 1949, Suppl. #3.

113 *The Canadian Woodworker*, Vol. I, No. 10, p. 1; *The International Woodworker*, April 13, 1949, p. 7.

114 *The Canadian Woodworker*, Vol. I, No. 13, p. 1.

115 *The Union Woodworker*, Vol. I, No. 21, p. 3.

116 *The Union Woodworker*, Vol. I, No. 18, p. 1; Vol. I, No. 21, p. 1.

117 *The Union Woodworker*, Vol. I, No. 26, p. 1.

118 *The Union Woodworker*, Vol. I, No. 27, p. 3.

119 Abella, *Nationalism*, p. 121, 128; MacNeil, *The IWA*.

Chapter 6

1 *Woodworker*, April 21, 1943, p. 1; September 22, 1943, p. 2.

2 Ralph Nelson, Interview by Jerry Lembcke, October 1977; *Monthly Labor Review*, October 1943, p. 727.

3 *Woodworker*, July 28, 1943, p. 1; March 4, 1942, p. 2; *Monthly Labor Review*, May 1944, p. 1031.

4 *Woodworker*, February 11, 1941; February 18, 1942.

5 *Ibid.*, December 16, 1942, p. 1.

6 Greeley to J.P. Boyd, reprinted in the *Woodworker*, March 17, 1943, p. 1.

7 *Woodworker*, March 10, 1943, p. 3.

8 *Ibid.*, December 1, 1943, pp. 1-2; November 24, 1943, p. 4.

9 *Monthly Labor Review*, November 1942, p. 967; September 1943, p. 522. March 1947, p. 414; February 1945, p. 316.

10 U.S. Bureau of the Census, Census of Manufactures 1939, 1947, 1954 (Washington, D.C.: U.S. Government Printing Office).

11 International Woodworkers of America, *Proceedings of District Council No. 3, 1944–47* (Collection at the University of Washington Library). See page 19 of the

1944 proceedings and pages 2-3, 11-25 of the 1945 proceedings.

12 International Woodworkers of America, *Proceedings of the Twelfth Constitutional Convention* (Portland, Oregon, October 11-15, 1948), pp. 302-304. International Woodworkers of America, *Proceedings of the Thirteenth Annual Constitutional Convention* (Vancouver, British Columbia, September 26-30, 1949), p. 78, 79.

13 Hartung, "Unpublished Memoirs," p. 50.

14 Al Hartung recalled the incident this way: "It was said that when Brown was working for the IWA, he and Fadling had played around some and it is said Fadling threatened to tell Brown's wife. Whether that was the reason or not, the fact was, Brown decided he would not run. So here were those who wanted to get rid of Fadling with no candidate. I was called into a meeting by Germer, Harvey Nelson and others and they said they wanted me to run for First Vice President. At first I thought they wanted me on the ticket with Brown but that was not the way it was. They said that Brown had pulled out and that Fadling had threatened to expose Brown if anyone ran against him" (Al Hartung, "Unpublished Memoirs," p. 43.)

15 *Ibid.*

16 International Woodworkers of America, *Proceedings of the Eleventh Constitutional Convention* (St. Louis, Missouri, August 26-29, 1947), p. 166.

17 The balloting procedure had been changed in 1950. Under the new procedure, after the balloting committee had completed its work and sent out its report, there was a ten-day waiting period during which a local union could protest the committee's results. After that, the International executive board was called in to take up local complaints and certify the election.

18 Hartung, "Unpublished Memoirs," p. 54.

19 Lyman Wax, Interview, October, 1978.

20 Hartung, "Unpublished Memoirs," p. 54.

21 Jack Greenall, *The IWA Fiasco* (Vancouver, B.C.: Progressive Workers Movement, 1965), p. 6; Karly Larsen, Interview with Jerry Lembcke, January 14, 1978.

22 "Minutes of 'Pro-CIO' Steering Committee Meeting, Vancouver, B.C., Tues., Sept. 26-49," p. 3.

23 Papers obtained under the Freedom of Information Act, see memo in Volume I, November 26, 1945.

24 *Ibid.*, August 27, 1946, August 18, 1951, January 18, 1946.

25 Karly Larsen, Interview, February 23, 1977. The harassment continued even after Larsen's work in the union had ended. In 1957, while confined to a wheelchair from an automobile accident, he was shot at with a .45 through a window; in 1957, the Larsen home was burned again.

26 The Smith Act, passed in June, 1940, provided for prosecution for "teaching and advocating the overthrow of the United States Government by force and violence" and for conspiring to do so. In 1940 it was used to force the finger-printing and registration of 3,600,000 non-citizen, foreign-born persons. As a consequence, the Party restricted membership to U.S. citizens between 1940 and 1944, which "cost the party about 4,000 members and substantially weakened its influence among the foreign born." See Foster, *History of the Communist Party*, p. 393.

27 The "first round" of the Smith Act trials was the Foley Square trial of eleven members of the Communist Party's national board. Arrested on July 20, 1948, were William Z. Foster, national chairman; Eugene Dennis, general secretary; Henry Winston, organization secretary; John B. Williamson, labor secretary; Jacob Stachel, education secretary; Robert G. Thompson, chairman of the New York district; Benjamin J. Davis, Jr., New York City councilman; John Gates, editor of the *Daily Worker*: Irving Potash, manager of the Joint Council of the Fur Workers Union; Gilbert Green, chairman of the Illinois district; Carl Winter, chairman of

the Michigan district; and Gus Hall, chairman of the Ohio district. Foster's case was tried separately because of his health (Foster, *History of the Communist Party*, p. 509). The second round of the Smith Act trials was the arrest in 1952 of "second echelon" Party leadership, i.e., district and state level leadership. The trial in Seattle was one of these. On trial with Larsen were Henry Huff, Washington state chairman of the Communist Party; Terry Pettus, Northwest editor of the *People's World* and a member of the Party's district committee. John Daschbach, chairman of the state Civil Rights Congress and a member of the Communist review commission; Paul Bowen, state leader of the Communist Negro movement and a district committee member; and Mrs. Barbara Hartle, former waitress and organization secretary for the Communist Party's Northwest District. Larsen was acquitted and the latter five were convicted. One defendant, William J. Pennock, committed suicide in the course of the trial. Pennock was president of the Washington Pension Union (*Seattle Post-Intelligencer*, August 4 and October 11, 1953).

28 It is also likely that the FBI found willing informants in the International office. A clue to this fact can be found in papers obtained under the Freedom of Information Act, memo in Larsen's papers, Volume II, March 30, 1953.

29 Transcript of trial proceedings, Volume XX, pp. 3186-3203 (Transcript in the possession of the authors).

30 *Ibid.*, Vol. XXII, pp. 3586-3590.

31 *Ibid.*, Vol. XXIX, pp. 4594-4604.

32 *Ibid.*, Vol. XXIX, pp. 4686-4687.

33 *Seattle Post-Intelligencer*, April 29, 1953.

34 *Ibid.*

35 *Seattle Post-Intelligencer*, October 8, 1953.

36 *Everett Herald*, May 18, 1953.

37 Al Hartung, "Unpublished Memoirs (in the possession of the authors).

38 International Woodworkers of America, *Proceedings of the Seventeenth Annual Constitutional Convention* (Vancouver, B.C., October 12-16, 1953), pp. 61-72.

39 Committee on Un-American Activities, House of Representatives, "Investigations of Communist Activities in The Pacific Northwest Area — Part 5 (Seattle)," Eighty-Third Congress, Second Session, June 16, 1954, pp. 6366-70.

40 Karly Larsen, Interview, January 14, 1978. The Communist Party's self-criticism was made in an internal Party document, "The Seattle Smith Act Case" signed by the Washington State CP Executive Board. Copy in possession of the authors.

41 Ed Kenney, Interview, December 1, 1976.

42 *Ibid.*

43 International Woodworkers of America Local 1-217, "An Analysis of Mr. E.W. Kenney's Preliminary Report on Resolution 26." (Ms. dated August 1, 1957; copy in the possession of the authors.)

44 International Woodworkers of America, *Proceedings of the Eighteenth and Nineteenth Annual Constitutional Convention* (Milwaukee, Wisconsin, August 22-26, 1955), pp. 154-157, 178, 243.

45 Ralph Nelson, Interview, September 23, 1976.

46 Tattam "Sawmill Workers," p. 78.

47 The "commie question" was raised throughout the entire period. In September of 1958, six months after the Special Convention, Hartung wrote to Germer, "The commies are raising hell up in Local 1-217. They have run somebody against Lloyd Whalen and several of the other offices. . . . We are doing what we can to help out but it is quite a complex situation. . ." (Hartung to Germer, September 30, 1958, in the Adolph Germer papers).

The CRDC politics of that period and even the present are difficult to categorize. They bear a greater resemblance to the Lovestone faction than to the social democrats of the CIO.

At the 1954 British Columbia District

Convention, Hartung said, "There are rumblings that some of our members in this district want to withdraw from the international union. This talk of revolt and secession must be making sweet music in the ears of management and the Communists, who tried so hard to smash the IWA a few years ago." Hartung said that "the secession move centered around discontent over financing of recent strikes in the southern and northern interior of B.C." (*Vancouver Sun*, February 17, 1954).

At the 1958 Special Convention it was also rumored that "Commies" were threatening to walk out. See International Woodworkers of America, *Proceedings of the Special Convention* (Portland, Oregon, March 24-28, 1958), p. 37; IWA, *Proceedings of the Eighteenth and Nineteenth...Convention*, p. 250.

48 IWA, *Proceedings of the Eighteenth and Nineteenth...Convention*, p. 481. Kenney says of that proposal, "We agreed theoretically that might be the best way to go but we thought we had to take it in steps... We felt that proposal came in from people trying to counteract the proposal from Sullivan (Resolution 26). The discussion that we had was on the basis that this would not be possible of achievement this time. It would be better to see where we might go with an alternative. But I know that in the separate caucuses that some of these people had, it was a feeling that that was a resolution put in there to scuttle this thing. At least that was the feeling...Ladd was sold on it himself...and the fact that it came in from way back there took it out of the hands of some people from Washington and Coos Bay" (Interview with Jerry Lembcke, December 1, 1976); IWA *Proceedings of the Eighteenth and Nineteenth...Convention*, p. 521.

49 The primary documents on Kenney's work are: E.W. Kenney, *Report of Resolution 26 Study* (Portland, Ore.: International Woodworkers of America, 1957);

IWA District Council No. 4, "Discussion Conference on Preliminary Report on E.W. Kenney Dealing with Resolution #26" (undated ms.; copy in the possession of the authors); IWA Local 1-217, "Kenney's Preliminary Report on Resolution 26."

50 IWA Local 1-217, p. 29.

51 *Ibid.*, p. 25.

52 International Woodworkers of America, *Proceedings of the Twentieth Annual Constitutional Convention* (Portland, Oregon, September 9-13, 1957) pp. 14-48.

53 *Ibid.*, pp. 143-45, 431.

54 *Ibid.*, pp. 157-161, 409.

55 IWA, *Proceedings of the Special Convention*, pp. 14-48.

56 *Ibid.*, p. 80.

57 *Ibid.*, p. 90.

58 Ed Kenney, Interview, December 1, 1976.

59 It was charged by critics of Kenney's report that his proposals smacked of AFL craft unionism and the influence of the business community. See the analysis by Local 1-217 cited above.

Chapter 7

1 Local 1-217, IWA, "Data on Lumber Industry," February, 1957. Figures on membership are taken from convention proceedings for the given years.

2 James Overton, "Uneven Regional Development in Canada: The Case of Newfoundland" in *The Review of Radical Political Economics* Volume 10 Number 3, Fall 1978, p. 111.

3 The introductory sketch draws primarily on interviews with Donna Butt, Ric Boland and Rolf Hattenhauer by William Tattam, June and July, 1978, St. John's, Newfoundland. Both Donna Butt and Ric Boland were directly involved in gathering the historical materials for the Newfoundland Mummer's Troupe Play about the 1959 IWA strike. Hattenhauer has been involved in extensive research on New-

foundland Labor History at Memorial University, St. John's, Newfoundland. For AND see the Financial Post Corporation Service Report on "Anglo-Newfoundland Development Company," September 13, 1956, pp. 1-4; this report along with a good deal of valuable information on organizing tactics and conditions both before and during the strike is contained in collected papers and documents at the IWA District Two Office, Toronto, Canada. Hereafter "cited" as IWA Newfoundland Papers.

4 Interview with Harvey Ladd, by William Tattam, 1978 at Burlington, Ontario.

5 *Ibid.*, Hattenhauer, "A Brief Labour History of Newfoundland," pp. 199-200; *Woodworker*, March 27, 1957, p. 1.

6 Interview with Joseph Smallwood, July, 1978, by William Tattam, St. John's, Newfoundland. In 1959, Mr. Smallwood was Premier of Newfoundland. Interview with Harold Pritchett. Ladd admits to the potential impact a successful strike conclusion would have on his IWA career, but stops short of making a direct connection to the IWA presidency. Interview with Harvey Ladd. See "Representatives Weekly Reports," by Newfoundland IWA organizers, Jeff Hall and Hank Skinner. IWA Newfoundland Papers.

7 International Woodworkers of America, District Two, *Press Release*, May 7, 1958. IWA Newfoundland Papers.

8 Copies of the Conciliations Board's Report are in the IWA Newfoundland Papers. For the AND response to the Conciliation Report and a general company history through the events of March, 1959, see "Turmoil In The Woods: A Report On The Dispute Between The International Woodworkers of America and the Anglo-Newfoundland Development Company," (Grand Falls, Newfoundland, AND Company, March 10, 1959), 31 pages.

9 Gwyn, *Smallwood*, p. 203.

10 Interview with Harvey Ladd.

11 Lush to Botkin, December 31, 1958. IWA Newfoundland Papers.

12 Hattenhauer, "A Brief Labour History...," pp. 207-210.

13 *Grand Falls Advertizer*, December 31, 1958; January 7, 1959. Reprints of editorials. IWA Newfoundland Papers.

14 See documents and financial statements from the Newfoundland Logger's Welfare Fund, IWA Newfoundland Papers; Harvey Ladd, "Statement over Newfoundland Regional Network of CBC," January 17, 1959. (Typed copy) IWA Newfoundland Papers.

15 The best source for background on Smallwood is Gwyn's *Smallwood*.

16 Tape of Smallwood's speech on CBC, February 12, 1959, St. John's, Newfoundland. Compliments of Ric Boland. In author's possession.

17 *Ibid.*

18 Commentary on the events of February 26 rely on interviews with Ladd, Smallwood and with Stirling Thomas, by William Tattam, July, 1978, Grand Falls, Newfoundland. Thomas was a mill worker at the Grand Falls' AND plant, became an executive officer in the NBWW, and later returned to the mill as a security officer. Gwyn, *Smallwood,* pp. 207-208: Williams, "Statement on Newfoundland," pp. 8-11.

19 Newfoundland, House of Assembly, *Special Session*, February 23, March 4, 5, 6, 9, and 10, 1959. Other sources: Donald McDonald, Secretary-Treasurer of the Canadian Labour Congress to Affiliated Unions, April 10, 1959, three pages; CLC, *Press Release*, March 12, 1959, three pages. Stanley Knowles, Executive Vice-President, CBC "The Facts About Newfoundland — A reply to Mr. Smallwood," (Ottawa, April 14, 1959), fourteen pages. CLC, *Petition to The Governor General in Council for Disallowance of the Newfoundland Trade Union (Emergency Provisions) Act of 1959*. (Ottawa, March 10, 1959), twenty pages. All in IWA Newfoundland Papers. Gwyn, *Smallwood*, pp. 209-211.

20 Hattenhauer, "A Short Labour History," pp. 215-216.

21 Leslie R. Curtis, Attorney General of Newfoundland, "An Address to the 88th Annual Meeting of the Canadian Manufacturers' Association," (St. Andrews, New Brunswick, June 10, 1959) p. 3. IWA Newfoundland Papers. See also House of Assembly, *Special Session*, March 6, 1959, pp. 7-8.

22 Interview with Tom Cahill, July 7, 1978, by William Tattam, St. John's, Newfoundland. Cahill was a CBC reporter at the time of the strike. Hattenhauer, "A Short Labour History," p. 216.

23 Joseph Smallwood, *I Choose Canada. The Memoirs of The Honorable Joseph "Joey" Smallwood* (Toronto: MacMillan, 1973), p. 406.

24 Material for the incident at Badger, Newfoundland, was drawn from Tattam's interviews with James Green and Danny Hiscock, plus interviews with Walter Paul and Oliver Lush, Badger, Newfoundland, July, 1978. Lush was Secretary-Treasurer of II, 251, the Badger IWA local. Walter Paul, a well-known Badger logger, was present at the incident. See also Hodgson "Reports from Newfoundland for the IWA International Executive Board," March 10, 11 and 15; Ray Timson, a reporter for the *Toronto Daily Star*, and an eye witness to the event, filed a four-page account, March 11, 1959. IWA Newfoundland Papers. And *Regina vs. Clifford Laing, Walter Paul, Uvan Roberts, Ralph Rideout, Augustus Colbourne, Archibald Colbourne, Ishmael Peddle, Augustus Rideout.* Deposition, Statements, etc., re. Riot Charges (1959). 148 pages. IWA Strike File.

25 Newfoundland, House of Assembly, *Afternoon Session*, March 11, 1959, p. 5. See House of Assembly, sessions on March 12, 13, 18-20, April 2-3, 1959, for Smallwood's statements and government actions following the Badger incident and the death of Constable Moss. For the events surrounding Moss' funeral and the consequent attacks on the IWA see Stuart Hodgson's, "Report," March 15, pp. 1-7; Hank Skinner's "Representative's Weekly Reports," March 11-14, 1959 and IWA, *International Executive Board, Afternoon Session*, March 17, 1959, pp. 1-14. IWA Newfoundland Papers. Interviews with James Green, IWA attorney, and Oliver Lush, who were inside the IWA office when it was attacked, were especially useful, as were interviews with Reginald Davis and Walter Tucker. Davis was president of the Royal Canadian Legion, Branch 12, Grand Falls, and headed the escort of Moss' body from the hospital to the train station. Tucker was an assistant purchasing agent for AND and was commander of the Grand Falls Militia. See also "History of Canadian Labour, Part XVIII," in *Textile Labor* (Textile Workers Union of America, AFL-CIO, CLC, June, 1974), pp. 10-13, for a general strike history as well as helpful details on the Badger incident and the aftermath.

26 Hattenhauer, "A Short Labour History," pp. 127-218; *The Globe and Mail* (Toronto) March 30, 1959, p. 9. Interview with Harvey Ladd.

27 Ladd to Joy Maines, Canadian Association of Social Workers, April 16, 1959. IWA Newfoundland Papers. Copy of the NBWW Constitution in the IWA Strike File.

28 Interview with Gonzo Gillingham and Walter Scott.

29 *International Woodworker*, June 8, 1949, p. 7; September 14, 1949.

30 *Laurel Leader-Call*, November 14, 1952, Laurel Collection.

31 Robert Hess to company employees, September 26, 1952. The letter was printed in the *Leader-Call* on the same day, Laurel Collection.

32 United Brotherhood of Carpenters and Joiners Leaflet, Laurel Collection.

33 *International Woodworker*, December 5, 1952; April 8, 1953.

34 Masonite Corporation to Employees, May 5, 1966, Laurel Collection.

35 Robert Analavage, "A New Movement in the White South;" Analavage, "Laurel Strike is Broken," Southern Patriot, 26 (1), January, 1968.

36 Jack Nelson, "Terror in Mississippi," *New South: A Quarterly Review of Southern Affairs*, Fall 1968.

37 Analavage, "Workers Strike Back," *Southern Patriot*.

38 Analavage, "A New Movement in the White South;" "Laurel Strike is Broken," *Southern Patriot*.

39 *Ibid.*, "Laurel Strike is Broken."

40 *Ibid.*

41 *Ibid.*

42 *Ibid.*

43 *Ibid.*; Claude Ramsay to Ron Roley, January 8, 1968, Laurel Collection.

44 Analavage, "Laurel Strike is Broken."

45 Ron Roley to Herman Blackwell, February 12, 1968, Laurel Collection.

46 Jim Youngdahl to Al Hartung, July 17, 1967, Laurel Collection.

47 Analavage, "A New Movement in the White South."

48 J. Baughman to Al Hartung, September 29, 1967.

49 Analavage, "Laurel Strike is Broken."

50 Dorothy Zellner, "Fifth Circuit Court Hears Laurel Suit," *Southern Patriot*.

51 J.D. Jolly to Ron Roley, March 29, 1968.

52 Analavage, "Laurel Strike is Broken."

53 Ray Craft to Ron Roley, March 4, 1968.

54 U.S. Court of Appeals, for the Fifth Circuit, J.D. Jolly, *et al.* vs. Walter Gorman, M.D.

Bibliography

Selected Bibliography

Articles, Books and Other Publications

Abella, Irving. *Nationalism, Communism, and Canadian Labour.* Toronto: University of Toronto, 1973.
Analavage, Robert. Laurel Strike is Broken. *Southern Patriot* 26 (1) January, 1968, 1968. A New Movement in the White South. *Southern Patriot,* n.d.
Aronowitz, Stanley. *False Promises.* New York: McGraw-Hill, 1973.
Bergren, Myrtle. *Tough Timber.* Toronto: Progress Books, 1966.
Bernstein, Irving. *Turbulent Years.* Boston: Houghton Mifflin, 1970.
Boyer, Richard and Herbert Morais. *Labor's Untold Story.* New York: United Electrical, Radio and Machine Workers of America, 1973.
Brody, David. *Steelworkers in America.* New York: Harper, 1960.
Chamber of Commerce of the United States. *Communists Within the Labor Movement.* Washington: Chamber of Commerce of the United States, 1947.
Chaplin, Ralph. *The Centralia Conspiracy.* Chicago: Charles H. Kerr, 1973.
Christie, Robert. *Empire in Wood.* Ithaca: New York State School of Industrial and Labor Relations, 1956.
Cochrane, Ben and Wm. Coldiron. *Disillusion, a Story of the Labor Struggles in the Western Woodworking Mills.* Portland: Binford and Mort, 1939.
Cox, Thomas. *Mills and Markets: A History of the Pacific Coast Lumber Industry to 1900.* Seattle: University of Washington Press, 1974.
Decaux, Len. *Labor Radical.* Boston: Beacon Press, 1970.

Dubofsky, Melvyn. *Industrialism and the American Worker*. New York: Vintage, 1975. *We Shall Be All*. New York: Quadrangle, 1969.

Egolf, Jeremy. *The Limits of Shop Floor Struggle: Workers and the Bedaux System at Willapa Harbour Lumber Mills, 1933-1935*. Manuscript, 1982.

Fonor, Philip. *The Industrial Workers of the World, 1905-1917*. New York: New World Paperbacks, 1965.

Foster, William Z. *American Trade Unionism*. New York: International Publishers, 1947.

Foster, William Z. *History of the Communist Party of the United States*. New York: International Publishers, 1952.

Galenson, Walter. *The CIO Challenge to the AFL*. Cambridge: Harvard University Press, 1960.

Gedicks, Al. The Social Origins of Radicalism Among Finnish Immigrants in Midwest Mining Communities. *Review of Radical Political Economics* 8 (3), Fall, 1976.

Glock, Margaret S. *Collective Bargaining in the Pacific Northwest Lumber Industry*. Berkeley: Institute of Industrial Relations, 1955.

Goodman, Irvin and John Caughlan. *The Truth in the Case of Laura Law, Murdered Labor Leader*. Manuscript, 1950.

Greenall, Jack. *The IWA Fiasco*. Vancouver: Progressive Workers Movement, 1965.

Gutman, Herbert. *Work, Culture and Society*. New York: Vintage, 1977.

Gwyn, Richard. *Smallwood, the Unlikely Revolutionary*. Toronto: McClelland and Stewart, 1968.

Horowitz, Gad. *Canadian Labour in Politics*. Toronto: University of Toronto, 1968.

Jensen, Vernon. *Lumber and Labor*. New York: Farrar and Rinehart, 1945.

Kampelman, Max. *The Communist Party vs. the CIO: A Study in Power Politics*. New York: Praeger, 1956.

Keeran, Roger. *The Communist Party and the Auto Workers Union*. Bloomington: Indiana University Press, 1980.

Kuczyniski, Jurgen. *The Rise of the Working Class*. New York: World University Library, 1967.

Kushner, Sam. *Long Road to Delano*. New York: International Publishers, 1976.

Larrowe, Charles. *Harry Bridges: The Rise and Fall of Radical Labor in the United States*. New York: Lawrence Hill, 1972.

Lembcke, Jerry. *The International Woodworkers in British Columbia, 1942-1951*. *Labour/Le Travailleur* 6 (6), Autumn, 1980. Capital and Labor in the Pacific Northwest Forest Products Industry. *Humboldt Journal of Social Relations, Spring-Summer, 1976*.

Levenstein, Harvey. *Communism, Anticommunism, and the CIO*. Westport: Greenwood Press, 1981.

Linberg, John. *Background of Swedish Emigration in the United States*. Minneapolis: University of Minnesota Press, 1930.

Lozovsky, A. *Marx and the Trade Unions*. New York: International Publishers, 1935.

MacNeil, Grant. *The IWA in British Columbia.* Vancouver: Regional Council No. 1 International Woodworkers of America, 1971.

Menefee, Selden. Tacoma, Timber and Tear Gas. *Nation,* July 17, 1935.

Midhailov, B.Y. et. al. *Recent History of the Labor Movement in the United States.* Moscow: Progress, 1977.

Morgan, Murray. *The Last Wilderness.* Seattle: University of Washington Press, 1955.

National Lawyers Guild. *Report of the Civil Liberties Committee.* Portland, Oregon, 1938.

Nelson, Jack. Terror in Mississippi. *New South: A Quarterly Review of Southern Affairs,* Fall, 1968.

Overton, James. Uneven Regional Development in Canada: The Case of Newfoundland. *Review of Radical Political Economics* 10 (3), Fall, 1978.

Parkin, Al. History of the IWA. *The B.C. Lumber Worker.* Serialized, 1947.

Parrish, Wayne and Wayne Weishaar. *Men Without Money, The Challenge of Barter and Script.* New York: G.P. Putman's Sons, 1933.

Phillips, Paul. *No Power Greater.* Vancouver: British Columbia Federation of Labour, 1967.

Pike, Robert. *Tall Trees, Tough Men.* New York: W.W. Norton & Company, 1967.

Prickett, James. *Some Aspects of the Communist Controversy in the CIO.* Science and Society 33 (3), Summer-Fall, 1969.

Rosenblum, Gerald. *Immigrant Workers.* New York: Basic Books, 1973.

Schlesinger, Arthur. *The Crisis of the Old Order, 1919–1933.* Boston: Houghton-Mifflin Company, 1957.

Schwantes, Carlos. *Radical Heritage: Labor, Socialism, and Reform in Washington and British Columbia, 1885–1917.* Seattle: University of Washington Press, 1979.

Scott, Jack. *Plunderbund and Proletariat.* Vancouver: New Star Books, 1975.

Starobin, Joseph. *American Communism in Crisis.* Berkeley: University of California Press, 1975.

Stephenson, Isaac. *Recollections of a Long Life.* Chicago: Isaac Stephenson, 1915.

Sparks, N. The Northwest General Lumber Strike. *The Comunist,* September, 1935.

Todes, Charlotte. *Labor and Lumber.* New York: International Publishers, 1931.

Tyler, Robert. *Rebels of the Woods: The IWW in the Pacific Northwest.* Eugene: University of Oregon Books, 1967.

Weatherwax, Clara. *Marching, Marching.* New York: John Day, 1935.

Weinstein, James. *Ambiguous Legacy: The Left in American Politics.* New York: New Viewpoints, 1975.

Yellen, Samuel. *American Labor Struggles.* New York: S.A. Russell, 1956.

Theses, Dissertations and Manuscripts

Barnhardt, Debra. *We Knew Different: The Michigan Timber Workers' Strike of 1937.* M.A. Thesis, Wayne State University, 1977.

Buechel, Henry. *Labor Relations in the West Coast Lumber Industry.* M.A. Thesis, State College of Washington, 1935.

Cary, Lorin L. *Adolph Germer: From Labor Agitator to Labor Professional.* Ph.D. Dissertation, University of Wisconsin, 1968.

Communist Party, Washington State. *The Seattle Smith Act Case,* n.d.

Crawford, Blaine. *Activities of the CIO and AFL in the Pacific Northwest Lumber Industry, 1935-40.* M.A. Thesis, University of Idaho, 1942.

Dunn, Marvin. *Kinship and Class: A Study of the Weyerhaeuser Family.* Ph.D. Dissertation, University of Oregon, 1977.

Gilstrap, Carlie May. *A Study of the Labor Dispute in the Various Decisions Involved in the Controversy at the Plylock Plant and the M&M Woodworking Company.* M.A. Thesis, University of Oregon, 1940.

Graham, Thomas. *Administration of Unemployment Relief in Oregon, 1930-36.* B.A. Thesis, Reed College, 1936.

Hattenhauer, Rolf. *A Brief Labour History of Newfoundland.* Manuscript, 1970.

Holsinger, Paul. *Patriotism and the Curbing of Oregon's Radicals, 1919-1937.* Manuscript, 1977.

Jensen, Vernon. *Labor Relations in the Douglas Fir Lumber Industry.* Ph.D. Dissertation, University of California, 1939.

Lembcke, Jerry. *The International Woodworkers of America: An Internal Comparative Study of Two Regions.* Ph.D. Dissertation, University of Oregon, 1978.

Lembcke, Jerry. *Labor Radicalism and the State in the Pacific Northwest: The International Woodworkers of America (CIO).* Manuscript, 1978.

Lembcke, Jerry and Wm. Tattam. *On Trial: Adolph Germer or the IWA.* Manuscript, 1979.

Lembcke, Jerry and Wm. Tattam. *The International Woodworkers of America: A Challenge to Vernon Jensen's Thesis of Rank-and-File Anti-Communism.* Manuscript, 1979.

Morris, Joe. *Communism in the Trade Unions.* Manuscript, 1965.

Nichols, Claude W. *Brotherhood in the Woods: The Loyal Legion of Loggers and Lumbermen, A Twenty Year Attempt at Industrial Cooperation.* Ph.D. Dissertation, University of Oregon, 1959.

Parks, Paul William. *Labor Relations in the Grays Harbor Lumber Industry.* M.A. Thesis, University of Washington, 1948.

Pritchett, Harold. *Maillardville-Coquitlam Ratepayers.* Manuscript, 1978.

Scribner, Tom. *Lumberjack.* Manuscript, 1966.

Smyth, L.T. *The Lumber and Sawmill Worker's Union in British Columbia.* M.A. Thesis, University of Washington, 1937.

Stone, David. *The IWA: The Red Bloc & White Bloc.* Manuscript, Simon Fraser University, 1973.

Tattam, William. *Sawmill Workers and Radicalism.* M.A. Thesis, University of Oregon, 1970.

Thompson, Michael E. *The Challenge of Unionization: Pacific Northwest Lumber Workers During the Depression.* M.A. Thesis, Washington State University, 1968.

Weinstein, Jacob Joseph. *The Jurisdictional Dispute in the Northwest Lumber Industry with Particular Reference to Portland, Oregon, 1937-1938.* B.A. Thesis, Reed College, 1939.
Whitten, Harry. *Subversive Activities: Their Extent and Meaning in Portland.* B.A. Thesis, Reed College, 1939.
Yorke, Dave. *The Workers' Unity League in B.C.* Manuscript, Simon Fraser University, n.d.

Oral History and Personal Memoirs

BC Overtime. Loggers Navy—History of the IWA in British Columbia. Vancouver: BC Overtime Educational Radio Productions, 1976.
Bedard, Jean-Marie. Interview with Wm. Tattam at Toronto, Ontario, 1978.
Ball, John. Interview with Jerry Lembcke at Gladstone, Oregon, 1976.
Boland, Ric. Interview with Wm. Tattam at St. John's, Newfoundland, 1978.
Brick, Mohr. Interview with Jerry Lembcke at Aberdeen, Washington, 1978.
Bridges, Harry. Interview with Wm. Tattam at San Francisco, California, 1968.
Butt, Donna. Interview with Wm. Tattam at St. John's, Newfoundland, 1978.
Cahill, Tom. Interview with Wm. Tattam at St. John's, Newfoundland, 1978.
Caughlan, John. Interview with Jerry Lembcke at Seattle, Washington, 1978.
Dalskog, Ernie. Interview with Frank Fuller, 1978.
Davis, Reginald. Interview with Wm. Tattam at St. John's, Newfoundland, 1978.
Fantz, James. Interview with Wm. Tattam at Portland, Oregon, 1968.
Foster, Ellery. Interview with Jerry Lembcke at Winona, Minnesota, 1981.
Gillingham, Gonzo. Interview with Wm. Tattam at St. John's, Newfoundland, 1978.
Green, James. Interview with Wm. Tattam at Badger, Newfoundland, 1978.
Harris, Bill. Interview with Jerry Lembcke at Reedsport, Oregon, 1976.
Hartung, Al. Interview with Wm. Tattam at Portland, Oregon, 1968.
Hartung, Al. Unpublished Memoirs.
Hattenhauer, Rolf. Interview with Wm. Tattam at St. John's, Newfoundland, 1978.
Helmick, Don. Interview with Wm. Tattam at Portland, Oregon, 1968.
Hiscock, Danny. Interview with Wm. Tattam at Badger, Newfoundland, 1978.
Kenney, Ed. Interview with Jerry Lembcke by telephone, 1976.
Ladd, Harvey. Interview with Wm. Tattam at Burlington, Ontario, 1978.
Larsen, Karly. Interview with Jerry Lembcke at Stanwood, Washington, 1977. Interview with Jerry Lembcke at Stanwood, Washington, 1978.
Lush, Oliver. Interview with Wm. Tattam at Badger, Newfoundland, 1978.
Nelson, Ralph. Interview with Jerry Lembcke at Portland, Oregon, 1976. Interview with Jerry Lembcke at Portland, Oregon, 1977.
Paul, Walter. Interview with Wm. Tattam at Badger, Newfoundland, 1978.
Paull, Irene. Interview with Linda Stiles at San Francisco, 1979.
Pilcher, Harry. Interview with Wm. Tattam at Portland, Oregon, 1968.
Pritchett, Harold. Interview with Wm. Tattam at Port Coquitlam, 1967. Interview with Harold Wynn at Port Coquitlam, 1970. Interview with Wm. Tattam at Port

Coquitlam, 1972. Interview with Simon Fraser University students at Simon Fraser University, 1975. Interview with Clay Perry at Port Coquitlam, 1978. Interview with Wm. Tattam and Jerry Lembcke at Port Coquitlam, 1978.

Rasmussen, Oliver. Interview with Wm. Tattam, 1978.

Roley, Ron. Interview with Jerry Lembcke at Oregon City, Oregon, 1977.

Ruuttila, Julia. Interview with Wm. Tattam at Portland, Oregon, 1968.

Scott, Walter. Interview with Wm. Tattam at St. John's, Newfoundland, 1978.

Smallwood, Joseph. Interview with Wm. Tattam at St. John's, Newfoundland, 1978.

Thomas, Stirling. Interview with Wm. Tattam at St. John's, Newfoundland, 1978.

Tomberg, Ernest. Interview with Wm. Tattam at Portland, Oregon, 1979.

Tucker, Walter. Interview with Wm. Tattam at St. John's, Newfoundland, 1978.

Wax, Lyman. Interview with Jerry Lembcke at Sheridan, Oregon, 1978.

Winn, Carl. Interview with Jerry Lembcke by telephone, 1982.

Archival Sources

Burns, Tom. Papers. Oregon Historical Collection, University of Oregon.

Germer, Adolph. Papers. State Historical Society of Wisconsin. Madison, Wisconsin.

International Woodworkers of America, Laurel Collection. International Office. Portland, Oregon.

International Woodworkers of America, Newfoundland. IWA Region Two Office. Papers. Toronto, Ontario.

Larsen, Karly. Papers. Personal Collection.

MacInnis, Angus. Papers. Special Collections, University of British Columbia.

Martin, Charles H. Papers. *Oregon Historical Society.* Portland, Oregon.

McAllister, Joseph B. Papers. *Oregon Historical Collection.* University of Oregon, Eugene, Oregon.

Pierce, Walter. Papers. *Oregon Historical Collection.* University of Oregon. Eugene, Oregon.

Portland Central Labor Council. Papers. *Oregon Historical Society.* Portland, Oregon.

Pritchett, Harold. Papers. University of Washington. Seattle, Washington.

Pritchett, Harold. Papers. *University of British Columbia.* Vancouver, British Columbia.

Sawmill and Timber Worker's District Councils Joint Executive Committee Conference. Papers. In author's possession.

Convention Proceedings

American Federation of Labor, 1937

Federation of Woodworkers, 1936-37

International Woodworkers of America, 1937-76

International Woodworkers of America, District Council No. 1, 1944-47.

International Woodworkers of America, District Council No. 1, 1948

United Brotherhood of Carpenters and Journers of America, 1936
Woodworkers Industrial Union of Canada, 1948

Union Executive Council Minutes and Proceedings

International Woodworkers of America. Minutes of the IWA Executive Board Meeting, January 11-12, 1940; Proceedings of International Executive Council, May 9-11, 1945; Special Executive Council Meeting, April 25, 1945; Proceedings of International Executive Board, August 2-5, 1945; International Executive Board Minutes, July 22-23, 1947; International Executive Board Minutes, August 21-30, 1947; International Executive Board Minutes, November 18, 1947; International Executive Board Proceedings, March 9-10, 1948; International Executive Board Proceedings, February 15, 1949.
International Woodworkers of America, District Council No. 1. Proceedings of Executive Board, March 7, 1945; Special Executive Council Meeting, May 20, 1945.

Union Documents

Congress of Industrial Organizations. Record of Hearing Conducted by CIO—appointed Committee of the Controversy Between Designated International Officers of the IWA and Adolph Germer, CIO-appointed IWA Organizational Director, 1941; Proceedings of Investigation Committee CIO, Re: District No. 1 and Division of Organization, IWA-CIO. Transcript, 1945.
International Woodworkers of America. Official Report of the International Tabulating Committee, 1946; The Voice of the International Woodworkers. Transcript, 1948; Minutes of White-Bloc Meeting, September 25. Transcript, 1949.
International Woodworkers of America, District Council No. 1. Untitled Trustees Report, June 21, 1948.
International Woodworkers of America Local 1-217. Date of Lumber Industry, 1957; An Analysis of Mr. E.W. Kenney's Preliminary Report on Resolution 26. In author's possession. 1957.
International Woodworkers of America, Local 1-80. Proceedings, Trial of John Ulinder. Transcript, 1945; Bulletin Centennial Edition, 1971.
Riddell, Stead, Graham & Hutchison. Report to the B.C. District Council, May 12, 1948.

Government Documents

State of Oregon. Unemployment Insurance Commission. *Report and Recommendations*, 1935.
U.S. Congress. House. *Hearings Before the Special Committee on Un-American Activities*. 76th Congress, 1st Session, 1939.
U.S. National Labor Relations Board. *Decisions and Orders*, 1939.
U.S. Congress. House. *Hearings before the Special Committee on Un-American Activities*. 83rd Congress, 2nd Session, 1954.

Newspapers

Aberdeen World, October 24, 1939.
B.C. Lumber Worker, 1931–1948.
Canadian Woodworker/Union Woodworker, 1948–1950.
Daily Peoples World, 1949.
Everett Herald, 1953.
The Herald, 1975.
Klamath Falls Evening Herald, 1939.
Labor New Dealer, 1937-1940.
Laurel Leader-Call, 1952.
Midwest Labor, 1938.
Militant, 1940; 1940a December 12:1; 1940b December 21:2.
Oregon Labor Press, 1933–1935.
Pacific Tribune, 1977.
Portland Oregon Journal, 1937–1941.
Portland Oregonian, 1930–1976.
Seattle Post-Intelligencer, 1953.
Textile Labor, 1974.
Timberworker, 1937–1940; June 11, 1937.
Timber Worker (Duluth), May 28, 1937.
Toronto Globe and Mail, 1959.
Vancouver Sun, 1954.
Voice of Action, 1933–35.

Index

221